RICHARD KERRIDGE

Richard Kerridge leads the MA in Creative
Writing at Bath Spa University. His essays have
been published in Granta and Poetry Review, and
he has twice won the BBC Wildlife Award for
Nature Writing.

RICHARD KERRIDGE

Cold Blood

Adventures with Reptiles and Amphibians

VINTAGE BOOKS
London

2 4 6 8 10 9 7 5 3 1

Vintage
20 Vauxhall Bridge Road,
London SW1V 2SA

Vintage is part of the Penguin Random House group of companies
whose addresses can be found at global.penguinrandomhouse.com

Copyright © Richard Kerridge 2014

Richard Kerridge has asserted his right under the Copyright, Designs
and Patents Act 1988 to be identified as the author of this work

First published in Vintage in 2015
First published in hardback by Chatto & Windus in 2014

www.vintage-books.co.uk

A CIP catalogue record for this book is available from the British Library

ISBN 9780099581390

Typeset by Palimpsest Book Production Ltd, Falkirk, Stirlingshire
Printed and bound by CPI Group (UK) Ltd, Croydon CR0 4YY

Penguin Random House is committed to a sustainable future for
our business, our readers and our planet. This book is made from
Forest Stewardship Council® certified paper.

For my mother and father

Contents

1	Palmate Newt	1
2	Common Toad	43
3	Common Frog, Marsh Frog, Edible Frog, Pool Frog, Smooth Newt, Slow Worm and Great Crested Newt	93
4	Common Lizard, Slow Worm, Sand Lizard	141
5	Grass Snake, Smooth Snake, Adder	201
6	Natterjack Toad, Aesculapian Snake	249
Further Reading		277
Acknowledgements		287

Chapter One
Palmate Newt

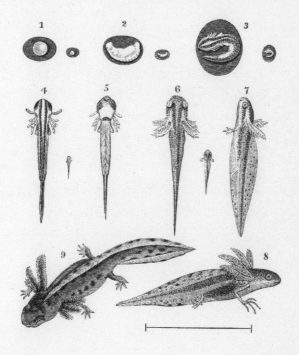

It started with a golden newt in a black bog. I was ten.

We were walking on Dartmoor, coming down a heathery slope. I can imagine the kind of day it was. Around us the moor looked beaten up by winter: muddy, uncommunicative, hunkered down into itself. The heather was faded, the bracken collapsed and papery, the grass washed-out, the earth wet. Cold breezes rushed at my face and found a way down the back of my neck. But the sun was out, and around me creatures were coming to life. Chirruping birdsong sprang up and was answered. Bumblebees wobbled into flight, to be snatched by the wind. Spiders scuttled from my feet. A crow called three times, and took off, flapping hard. The gorse bushes were dense with yellow flowers. Around them, in the sun, there was a coconutty warmth.

Our path dropped towards a boggy stream and a bridge of flat stones. Pools glittered there. I ran ahead and lay on the bridge to look into the water. Quietness settled around me. The April sun warmed my back. The stream was shallow here, hardly moving over deep-looking mud. On either side, there was bog, clumps of ribbony brown grass, rising out of black slime. Between them were dark shallow pools. Patches of oily film gleamed on the surface. Tiny beetles raced in circles, catching the light.

3

Beneath me, under the water, the sun picked out a soft brown landscape. Particles drifted and settled, coating half-buried leaves and sticks, and making little turrets that trembled. A beetle dashed jerkily about. On the bottom, a long-legged stick-like creature, some sort of insect larva, seemed to have become furred with algae. It struggled feebly. Repelled by it, I looked away quickly, and a movement nearer the middle caught my eye. A small animal, corkscrewing like an otter, swam to the surface and gulped out a bubble. As soon as it had gulped, it stuck out all four legs, and let itself drop, paddling once or twice on the way down. Tail and back feet reached the bottom, sending up a puff of mud. There the animal rested, half-floating. It was a newt.

But it wasn't like the pictures I had seen. This newt was smooth and light brown, about two inches long. I looked down at the head, densely speckled with gold. It seemed threaded with gold, faded tapestry gold. There was no wavy crest, but the top of the tail was a fine edge, crinkled in places. I leaned out to get more of a side view. The newt didn't move. On its sides it was mottled with green. The forefeet were long-fingered sensitive hands that stirred in the water and steadied the body. I couldn't see the back feet in the mud.

Afraid to take my eyes off the newt, I began to roll up my sleeve, thinking about reaching down into the water and manoeuvring my hand for a grab. But I knew the attempt would be almost hopeless. If only I hadn't left my net in the car. I was too high up from the water, and there was no other way I could get near. The bog was obviously too deep for wading. And here came my family, approaching on the path. They were talking. Frantically, I looked up and shooshed at them. 'What is it, Rich?' my dad called.

'A newt. There's a newt here.'

'Where?' He came and looked over my shoulder. 'Stay back,' he said to the girls.

It looked tiny now, and far off, a detail on the stream bed.

'Ah yes. I see it. A little brown newt. Interesting.'

'Can *I* see?' My sisters came onto the bridge, pressing behind us. 'Where is it? I can't see anything.'

'Follow my finger,' said Dad.

'Oh.' Cathy wrinkled her nose. 'It's just brown.'

'It's quite sweet,' said Anne.

'I suppose.'

'Dad,' I said. 'Can we go back to the car, for the net?'

'It's half an hour away.'

'Please,' I said. 'Please.'

'We're in the middle of our walk.'

'Please.'

'Where is it?' said my mother. 'Oh yes. There.' The girls had moved off the bridge and were playing some sort of skipping game.

'Please. I'll run all the way.'

'I've said no. We've got to have our picnic.'

'You have lunch while I run there and back.'

'You can't go on your own.' He looked uncertain. 'Come on. It's just a newt.' He took a few steps down the path. No one followed.

'Come on. Now.'

'Come on, Rich,' said my mum. 'It's better to leave it where it is. If you caught it, what would you do with it?'

'I'd take it home and keep it, in a tank, with rocks sticking out of the water, and plants. It would be my own zoo. Please. I've always wanted a zoo.'

Dad had come back to the bridge. 'I'm telling you, we're not going back for the net. Come now.' He paced up and down.

'It's still there,' said Cathy, on the bridge again.

'What would you call it?' said Anne.

'Please let me go back. I'll run all the way.'

'Don't be silly. You'd get lost. Now that's enough. We're walking on. Come now.'

And this time they followed him, leaving me on the bridge, gazing hopelessly at the brown creature, at a loss to explain why I wanted it so much. Not more than two inches long, it was huge in my mind. I wanted it in my own underwater forest, where it would emerge from a thicket of weed and come up against the glass, close to my face, in clear watery light.

I loved zoos. Every year on this Easter holiday we went to Paignton Zoo, where there were tigers behind glass. The outdoor part of their enclosure was an unmown lawn planted with stands of bamboo. There was a concrete pool, and at the back a rocky slope, with a cave at the top. Sparrows chattered in the wire netting roof. As I watched, I imagined coming upon that cave unawares, and hearing a growl from inside. In fact the cave led through to the indoor section, and if the tigers weren't out you could go round and see them behind bars.

The newt still hadn't moved. I turned and trotted after my family.

But where did it all start really?

Earlier in my childhood, I had discovered the wildlife of Africa and Asia, as seen in picture books and the tea-packet cards I collected. African wildlife lived on yellow plains. Giraffe browsed on tree-tops. The hippo's skin shone. So did the coat of a black sable antelope, kneeling to drink. A warthog ran, tail in the air. These creatures were bright in the sun. Indian wildlife seemed shyer, peering from shadowy places. The colours were strange. A huge antelope called a nilgai was blue. Against a mauve and yellow sunset, fruit bats hung from branches, their dark faces watching. These creatures looked wise and mysterious. All of them thrilled me.

Framed in a tiny card, the lion walked in disdainful profile, intent on his business. A tiger's head came up out of the bamboo forest, eyes narrowed. On a sandbank, a crocodile grinned. The red river hog had shocked eyes and white tassels on its ears. A mandrill slid its fingers beneath a stone. Stranger creatures looked out at me too. Their names were new to me, and at my school desk I would whisper them: tarsier, zorilla, Coke's hartebeest, gaur. The ratel was a brutal-looking badger. Serval and caracal were delicate lynxes. My eyes searched the backgrounds. I wanted to see further. Here was a blur of trees, there a shimmer of grass. For some people this landscape was background. It was normal. What did it conceal? I looked hard, trying to see around the animals. At their feet, there was scuffed dry earth and flattened grass, like English grass in summer. I thought of the touch of that earth and grass, remembering my bare feet on our lawn.

Later, I read *The Jungle Book* and *Jock of the Bushveldt*. On television, there was *On Safari*, with Armand and Michaela Denis. My mother's parents lived in two rooms in our house, and had a television long before we did, since my father did not want 'those jabbering idiot voices interrupting our thoughts'. Sitting at my grandparents' feet, I got to know the African bush in grainy black and white. I loved its texture. Over there, said Michaela, in an elegant hushed voice, was a group of lions resting under a tree. Three, she thought. Oh, and another. They were indistinct shapes. The camera went closer. A lioness, clearly visible now, got up and stretched her front legs. This programme taught me that a lion's roar at night was not as I had imagined it, and not the rippling snarl my mother had produced over my picture books. A true roar was a single, downward note, an abrupt release of breath.

I saw gazelles shimmer in the distance. Elephants flapped their ears. Buffalo were like a black sea. Most dramatic was the Black

Rhinoceros, with its heavy beaked head and the huge jutting spikes of its nose-horns. There was something terrifying about the way this animal moved. It was unwieldy, yet it exploded from the bushes so quickly, swerving and bucking, feet drumming the ground. Even on my grandparents' carpet, I felt too close. When a rhino's head emerged from the thornbushes, I cowered. In Africa, people lived in the same space as these creatures. They really did. There was no fence enclosing the animals. Rhino were free to walk everywhere, lions too. If you were walking, and one of these animals came out of the bush, it could cross the intervening space in a second, and be on you.

Britain had no wild animals like this. The nearest to it were the bulls we sometimes saw on our country walks. Some of them, looming above the fence, seemed almost rhino-like in size, and just as dangerous. I dreamed one had got into the garden and was roaming around the house, looking for a way in: a big black and white bull. On a walk I always hoped to see one.

But if you got close to the ground, and saw the forest and savannah down there, Britain did have animals as strange and beautiful and savage as the ones on my tea-packet cards. Small faces were there, looking up from the grass and leaf litter. Speckled creatures flared out in sudden colour. There was stalking, there was fighting; there were scurrying dashes for safety. Large, scarred old veterans would glide into sight, or bask lazily. Hordes would move through the grass stems, bringing a whole field to life.

This book tells the story of my discovery of Britain's amphibians and reptiles, the way they entered my imagination, and the adventures I had with them as I grew up. Perhaps they were odd creatures to take centre stage. My wildlife fascination might have settled, more conventionally, on birds. Sometimes I went bird-watching. At school there were boys who were keen. I was excited by the suddenness of birds, and the sweep of their movement in the air. But their elusiveness was frustrating. The space they lived in had no boundaries. I wanted drama with a stage, and animals I could creep up on and catch and possess. Birds of prey might well have captivated me, if there had been any near at hand. But in London in the 1960s there didn't seem to be any hawks, and in any case the idea of reaching out and catching one seemed fanciful. Years later I read with amazement that some boys at this time kept kestrels.

Large mammals might have held my interest, if there had been any in the London suburbs. The only ones were foxes. I heard them shriek at night. But they were rarely seen and quick to flee. Badgers or otters would have presented the same problem, if any had been there. These were not creatures I could get near. But amphibians and reptiles did not live separate, untouchable lives. They were wild and mysterious, yet small enough to be picked up and kept in enclosures, where they would not seem domesticated and too big for their surroundings – as a fox does, or an otter, at the zoo – but would melt into the simulated landscape and stay wild. These animals would not be tamed. They would not nuzzle at their owner's fingers.

And they might already be living in my garden, I was excited to learn. Frogs, toads and newts might be there. They were certainly in parks and wild patches nearby. They lived in my world, but made it seem larger and wilder. The thought of these animals made the smallest patch of wild ground full of promise. Suddenly, my garden and the park at the end of it no longer seemed to be enclosed spaces, thoroughly finite, whose possibilities could quickly be exhausted. They seemed part of an infinite space, a depth of wildness on all sides, receding into the distance. This depth was wild England in its immense completeness, held together across the cities by secret corridors: railway-banks, river-banks, wasteland. Frogs, toads and newts – just the thought of them – did this for me.

They were not an uncommon interest among city boys then. Children were routinely given frogspawn or tadpoles, and older boys often kept newts or Grass Snakes in tanks for a while, though few with the obsessive dedication that my friends and I developed and that bound us together. These were still the days when boys of eight climbed trees and dammed streams and lit campfires in the woods, dreaming that they were Robin Hood. Wild England for a while still held its own against the metallic surfaces and control panels of space ship and time machine. The collecting of wild birds' eggs had only recently been made illegal.

Some boys collected butterflies, chasing them with nets and forcing the fluttering insects into jars containing cotton-wool soaked in chloroform, before pinning them in neat rows on white card, identified by the English name and the Latin name and the symbol for male or female, in careful fountain pen. Wooden cabinets could be bought, though they looked expensive. These cabinets could turn your bedroom into a museum. They contained shallow glass cases that pulled out as drawers, in which to display your rows of butterflies, whose wildness would be strictly confined by the extreme order of the surroundings, but also given definition and concentration. Yet somehow the writing never looked quite neat enough. And beneath the velvet bodies mounds of dust began to appear, and oily spots on the paper. Wildness was defiant. I was a little attracted by this hobby but more powerfully repelled. The killing was horrible, and so was the lifeless order of those displays. I wanted my wildness captive and near at hand, but not dead and still.

Not many children catch and keep British reptiles and amphibians now. At least, that is my impression. Perhaps more do than I imagine, since it would be a largely invisible activity. Public culture, on television and in magazines and books, does not encourage the catching of wildlife, as once it did. The idea would now seem shocking. Gone are the books for young naturalists giving careful instructions for finding and capturing the animals, housing them, feeding and preventing escapes. In the 1960s, there were many. The rarer species are now strictly protected: the Sand Lizard, the Smooth Snake, the Great Crested Newt, the Natterjack Toad and the recently reintroduced Pool Frog. Without a special permit, it is illegal to disturb these animals or their habitats, though plenty of people do. No native species at all can be captured for sale. And most of the European species once found commonly in pet shops appear there no longer. Wild catching for the pet trade is now unlawful throughout western Europe.

Quite right too. Many animals were injured in transit, or died in the pet shops or in the hands of children with little idea of their needs. In the wild, most of these species are in decline. Habitat loss is the main reason, but collectors for the pet trade can all but wipe out a local population. I am glad that the laws have changed. But I am wistful for the big glossy Green Lizards, the quick Wall Lizards with mosaic markings, the gold-dusted European Pond Tortoises and viscous Tree Frogs, green and shiny as new leaves. Sometimes, stuck to the glass of the tank, the Tree Frogs were furled up tightly like sharp buds.

I am wistful, too, for the way our fascination with the native species – the few native species, in their subtle browns and greens and sudden radiance – took us to sandy heaths where we crouched watching, and ponds and dykes where we waded and slipped on the banks. Keeping reptiles remains popular. Of that there is no doubt. These animals continue to fascinate. They have sinister power and sexiness. Voldemort in the Harry Potter novels has a snake as his closest companion. Often the evil is erotic and hypnotic. The cover of the December 2013 edition of the style magazine *GQ* shows an image designed by the artist Damien Hirst, who is famous for his use of animal bodies. Rihanna, the pop icon, appears as a modern

Medusa, naked except for a large Boa Constrictor draped around her neck. The snake seems to be beginning to move down her body. Its tongue is flickering. There is a gothic tattoo beneath Rihanna's breasts. Her hair has been replaced by a tangle of snakes, their heads all waving. I can see that they are Corn Snakes, small Boas and Milk Snakes, the latter chosen for their resemblance to the deadly American Coral Snake. One Boa, rising from the top of Rihanna's head, has been digitally enhanced to give it a fanged gape like a viper's (Voldemort's snake is a similar hybrid in the Harry Potter films). Rihanna's eyeballs are green and have narrow black vertical pupils. In another shot, she too has long viperish fangs and is making a strike.

Newspapers have claimed that at Miami airport the smuggling in of exotic reptiles is in money terms second only to the smuggling in of drugs. Even rare British species are sometimes smuggled out for illicit trading. There is a subculture, a huge world of enthusiasts gazing into small tanks where animals bask under light bulbs. But nowadays, as any reptile shop shows, it is mainly a matter of tropical species in heated tanks: fat-tailed geckos, chameleons, Australian Bearded Lizards. American Corn Snakes and small pythons or boas are fed on defrosted 'pinkies' – hairless baby mice, sold by the dozen. Geckos and pythons are bred for rare colours and patterns. The language becomes technical. Brand names appear. Instead of earth or sand in the tanks, there is 'substrate', perhaps a designed product such as 'vermiculite' or 'lignocel' – absorbent, odour-free, digestible and without sharp pieces. Maximum functional efficiency is the aim, and the function is purified, abstracted from the messiness of ecological process in the wild. Somewhere, in all this, is still the romance of wild drama. In some of the tanks you see in rows in pet shops, it is possible to glimpse desert ravine or rainforest. But the vision is very elusive. Usually the confinement of the animals seems sad. They are so far from home. It is such poor simulation.

What makes me happy, in thinking about reptiles and amphibians now, is still what excited me at the age of eight. They make the world around me seem wild. I still feel this, driving or going by train through strange cities. Looking out at an overgrown bank or green ditch, or across puddled fields, I wonder what lives there. What would I see if I went close? I imagine it – the sensation that always comes when I am out looking for reptiles, running my eyes across a bank. Everything goes quiet. Background noises disappear. Winds no longer mutter at my ears. Thoughts cease to clamour. I feel, in my chest and my back, small sensations of falling.

On a hot June day last summer, I was in a train just outside Swansea. The train stopped. We waited several minutes. It didn't move. Faint, isolated sounds came from its works underneath us, creaking and easing sounds, as if its body was relaxing. I was sitting beside a window, and outside, close up against us, was a steep bank, made of chippings but overgrown. Brambles had forced their way through, and were sending out thick, thorny seekers. Here and there I saw feathery moss. My eye ambled on. Bees wandered around. A white butterfly flickered and settled. There was coarse grass in places.

My eye came to a part of a Grass Snake: just a part – a few inches of body.

It was all I could see. The head was concealed, and the tail. But I saw the familiar green-olive skin, with the little black flecks. Suddenly I was straining to see more. I pressed myself against the window, trying to find a new angle. Could I startle the snake into moving by tapping the glass? It might then reveal itself, but it would glide away, and there would only be bare chippings. Perhaps the seat behind mine would afford a better view; it might show me the head. I stood up. There was someone in the seat. I pretended to be stretching, and caught the same little view of the snake's back, from a slightly different angle, but could see nothing more. I felt delighted. The snake was an old friend, a visitor from childhood. For a moment I thought of announcing its presence to the carriage, calling everyone to cluster round and see. Surely they would be interested. But the train began to move. And it seemed strange to me that although the snake was liable to whip away in an instant, it was instead we who were moving and leaving the reptile behind, quite still, in the murmur and heat of the morning.

This book is about what these animals mean – what they meant to me in childhood, and what they mean now. It is also about the animals themselves, Britain's newts, frogs, toads, lizards and snakes. What is happening to them now, in a country where the human imprint is everywhere? Where and how can we watch them? What is their likely future?

And why did I love them? What other things were happening at the time? My family must come into this and my friendships. The passionate love of wild nature that sprang up for me was undoubtedly a response to other things in my life. Certainly it was a refuge, a place where I could shut things out. But, because I relaxed there, it was also a place where I could release feelings and see the shapes those feelings took – shapes that formed and then dissolved. Wild nature traditionally does this. But modern environmental problems teach us that wild nature should not only be seen as a separate zone and a place of escape. Large systems such as climate patterns and industrial practices connect what happens in wild nature with what happens in cities, and some natural wildness is to be found in the most built-on environments. For me, one of the joys brought by reptiles and amphibians was that they made the suburbs wild.

In the book, I jump around between various kinds of writing. Natural history information mixes with passionate description of these animals as I see them now. In and out of this material threads the story of my growing-up – a story of adventures with reptiles and amphibians and a broader story of friendship, family and change. There is a cultural and historical story too. What have these animals meant? All these things can be separate spaces; sometimes they *need* to be separate spaces. But I don't think they can be separate for long. Leisure surfaces in work, work in leisure. The place to which we escape is the place where we confront. This is a book about personal meaning – *my* personal meaning, in which events from the past are reconstructed according to my instinct, my prejudice and my need – and a book about public places, such as nature reserves, that anyone can enter.

All day, that newt in the Dartmoor stream moved around in my mind. I pestered Dad to take me back there.

'Please, Dad. I really want one. Can we go back tomorrow? I know I can catch one. Please.'

He was standing at the top of the steps leading down to the beach. A small plastic spade was held in his clasped hands behind his back, like an army officer's baton. Dad was reading a large notice.

NEVER BATHE ALONE

DO NOT BATHE IN HEAVY SURF

ALWAYS STAY WITHIN YOUR DEPTH

TAKE ALL YOUR LITTER HOME

DO NOT PLAY RADIOS ON THE BEACH

RESPECT OTHER BEACH USERS

'Keep your back straight,' said my dad. 'Don't pick your nose.'

'Da-ad!'

He turned and smiled.

'All right,' he said. 'All right. I'll take you tomorrow.'

This time I had my net and a large jar. The others had gone to the beach. Dad sat down with his newspaper and pipe. 'There you are, then. Half an hour.' Gripping the net, I crept onto the bridge and peered over, my eyes going straight to where the newt had been. I saw nothing. This was an overcast day, and the mud was darker. Wind blew ripples on the water and fluttered the bleached grass. There were prickles of rain in the air. I was scared it would start and my father would call me away.

Seeing nothing, I jabbed the net under the overhanging grass, making clouds of black mud billow out into the stream. I swooped the net along. In seconds it was heavy with mud, the weight bending the bamboo stick as I brought the net up onto the bridge, and peered in.

Something was moving. Another of those struggling insects. No, something shiny and plump. I hooked a finger under, and raised the creature out of the slime. Four legs, and a little dog-like head: it was a newt, smeared all over with mud. 'I've got one,' I called to my father.

'Jolly good.'

'I've got one,' I gasped in delight to myself. The newt was now walking with determination across the lump of mud in the net. With the net in one hand, held over the bridge, I managed to reach down with the other and half-fill the jar with water. Gently, between thumb and finger, I picked up the glistening newt and placed it on my palm. At the tip of the tail, what looked like a tiny hair dragged across my wet skin. I dropped the newt into the jar, where, instantly washed clean, it was dazed for a moment, then dived and darted like a fish, trying to burrow into the glass. I held up the jar.

COLD BLOOD

In the clean water, with no dark background, the colours were different, more greeny brown and minnow-like. The underside was white. On the belly was an oval of pale yellow. There was a fish-like semi-transparency. Pinkish shapes moved in the white throat. Down the white underside of an arm ran a thread, a pink vein. The back feet, much larger than the front and thickly webbed, seemed suffused with black ink. Behind them the tail was larger than the body; it was orange, marked with two lines of black blotches. At the end was that tiny black hair.

My father inspected the jar in silence. 'Do you want it, Rich?' he asked, after a moment. 'Do you really want to take it back to London?'

I had no doubt at all. A further half-hour's trawl under the rooty banks yielded one more newt, slightly smaller than the first. In a glow I walked back to the car, full of plans for my zoo.

Every spring we had a week at a small dairy farm near Kingskerswell. I loved driving the Guernsey cows down the lane, loved the sweet smell of them, loved the mud and straw and cow-splash of the yard, loved the crooning hens, loved the warm eggs in the nest-boxes, loved the old concrete bull-pen with its bent and rusty railings; there was no bull, but I imagined one. Meals were quite formal, in the large farmhouse dining-room. Four or five families were staying. Each had their own table. We had breakfast and dinner there. Lunch we got for ourselves, usually a picnic.

At the next table that evening, there was a girl of about my age, a tall girl with brown hair in a ponytail. Her name was Isobel; on the first day our families had made gruff introductions. She was reading one of the Narnia books. I was reading one of my war comics. In heavy black ink, on coarse paper, the German bombers were crossing the channel in massed formation. Up came the Spitfires to intercept them, but cruising high above the bombers was a squadron of mottled Messerschmitts, ready to swoop. 'Come into our trap, Englander,' purred *Fliegerleutnant* Kessel, smirking in his narrow cockpit. 'Just a little bit closer.' The German fighters waited. They flipped over and came diving out of the sun.

Flying Officer Peter Bradshaw, though secretly wearing contact lenses, managed to jerk his Spitfire up out of the line of fire. Juggling his mixture control, Bradshaw made the Spitfare cough and buck in the air, so that the Germans passed beneath him. His thumbs hovered, as the last one came into his sights. Then he pressed down, and the Messerschmitt's tail broke into pieces. The lenses were a new invention, developed by Bradshaw's old science teacher. 'Next time, Englander, you will not be so lucky,' fumed Kessel. The cruel Prussian was right, because Flight Lieutenant Cathcart, jealous of Bradshaw's success, was plotting to steal the lenses. I bought one of these comics every day of the holiday, walking on my own to the end-of-beach shop, where there were naughty postcards and little bags of plastic soldiers. My net came from that shop too.

Oxtail soup was placed in front of me, and I looked around the room. Against the wall near our table, tantalisingly near, was a glass-fronted bookcase, and in it, among cricket books, Second World War books and books about Dartmoor, I spotted *Animal Life of the British Isles*, an old brown book with gold lettering on the spine. 'Can I look at that?' I asked.

'I'll ask Mrs Rose after dinner,' said Mum. 'Eat your soup. Put that comic away.'

The book was heavy. I handled it gingerly. Glue was flaking off the spine. The author was Edward Step, F.L.S., and the title page said 'London and New York 1921'. Opposite that was the first illustration, a photograph captioned 'Hedgehog preparing to attack Grass Snake'. It looked as if the colour had been added by hand. Mossy greens in the background were drab yet unnaturally intense. A whiskery hedgehog was turning, led by his twitching nose, towards a grass snake coiled a few inches away. Rocks in the background were slightly blurred, and the animals looked posed and stiff, not quite real, like stuffed animals placed there. Yet the picture had an exciting wildness about it, just because of the oddness of colour and texture. The hedgehog had black twinkly eyes that peeped out through swirling grey hair swept back from the face – the look of a malevolent tramp. Half-covered, where the hair met the spines, was a grey but human-looking ear.

I looked through the book for more pictures. Some were black and white photographs, dingy and badly focused. Often the animal wasn't very clear. Others were coloured but too highly coloured, creating odd textures. Fur, especially, showed too much detail, standing up unnaturally or looking too bedraggled. A Serotine Bat, with a face like a calf, crawled towards me. The Long-eared Bat was a sinister blot of dark fur; I couldn't decipher the shape. It was a portal into blackness. 'Ears uncurling after sleep', said the caption. Leisler's Bat was a ball of fur from which a single tiny hand protruded, like the hand of a drowning swimmer. A Common Lizard looked emaciated and obviously dead, lying with its nose in the ground. The colouring had been done with a roughness that seemed brutal. Strange blue stones shone out in the gravelly foreground. Another, a black and white picture, showed two lizards dead on bare wood. They were arranged, it said, to show the different kinds of scales on back and belly. Each was held in place by a nail in the wood, between forearm and body. The Common Toad, in a dramatic picture, was a huge monster, puffed up and angry, glowering at something tiny on the earth. All these illustrations had the same murky, slightly unreal quality, and in them I saw an older, wilder England, full of gloomy woods, moors and swamps, and elemental creatures. The newts were at the very back of the book.

There was the Great Crested Newt, *Molge cristatus*, male, with a huge golden eye, and the Smooth Newt, *Molge vulgaris*, mottled like a Messerschmitt. 'Male in bridal attire', it said strangely, under the picture. 'The male seeks to excite the female,' said the book, 'by displaying his beautiful crest and his heightened colours; also by rubbing her with his head and lashing her with his tail.' Another picture of a male Smooth showed him falling back in the water, arms waving, as if struck. His eye was rolling. 'Male, underside', said the caption. The underside was bright orange spotted with black. Over the page was a female Smooth with an extra front leg, illustrating the ability of newts to regenerate lost limbs, or sometimes grow an extra limb from the gap where one is partially severed. The next illustration was a Palmate Newt, *Molge palmatus*, male, and this, unmistakably, was what my newts were, both male.

That hair or thorn at the end of the tail made me certain. It was clear in the picture. The book called it a thread. Apart from that, the newt in the illustration, seen from below, had few memorable features, its belly a watery yellow, its back a drab green. It floated in front of some spindly pondweed. There was little resemblance to the gold-encrusted brown creature I had seen from the bridge. The watercolour tones had the moody, interestingly melancholy character that pervaded the book, but this time there was no drama to energise those tones. My father had said, slightly puzzled, 'It's just a newt.' For a moment, looking at that picture, I might have falteringly said the same. The darker shade into which the background modulated at the top of the picture was meant to suggest the surface of the water. It looked like a rain-heavy sky.

But I rallied. The two newts were now in a large tin milk-can supplied by Mrs Rose, which I had filled to about six inches deep with stream-water. I knew they sometimes needed to come out on land, so I put in a tangle of weed rising above the water, and a snapped-off twig with several branches, which gave the appearance of a sunken tree in a swamp, perhaps in the Everglades or Congo. To complete the effect, I dug up a tussock of grass and put it in, roots and all. This made the newts very difficult to see, especially when I had just put the grass in and the mud hadn't settled. In the muddy swirl a gold-streaked head appeared, gulped and vanished, and I was thrilled again. The can stood on a bench in one of the barns, where Mrs Rose said I could keep it, and several times each morning and evening, I would creep up to the can, so as not scare the newts out of sight.

On our last day but one, I was on my way to see them before breakfast, when I saw Isobel standing by the gate. She had been collecting the eggs, something I liked to do too. The basket was on her arm. 'Hello,' she said, as I approached.

'Hello.'

'Mrs Rose says you've been catching newts.'

'I've got two Palmates,' I said. 'Palmate Newts.'

'Are they a special kind?'

'They are the species you find on Dartmoor,' I said.

'Can I see them?'

'Yes,' I said. 'OK.'

She had a voice that I knew was posh, but I couldn't tell how posh. I associated that kind of voice with large houses behind tall hedges, neat school uniforms, black labradors and schools that looked like stately homes; but also with a lot of the books I enjoyed – school stories, and the Narnia books themselves. My first sight of Isobel's family in the dining room had prompted me to straighten my back, eat nicely, pass dishes attentively and laugh politely at things said. I don't know why. Apparently people noticed. 'I heard that man say, there's a well brought-up boy,' my mother told me that night.

'You might not see much,' I said, as I led the way. 'They're only small.'

The newts seemed to have darkened. They were almost the colour of the mud at the bottom. Isobel knelt on a stool to look in. I pointed out the newts to her – one under the twig and the other half-under the weed. 'Oh yes,' she said. Then the twig newt came up for air, with a flash of the gold on its head, and of yellow as it shimmied back down. Isobel gasped with delight – really gasped. 'Oh, it's beautiful. A little dragon.'

'You actually like them?'

'I know I seem a prig,' she said abruptly. 'That's what people think.'

'Oh no,' I said, surprised by that word more than anything else. It was a word out of books; a word from Narnia. I'd never heard a real person say it. 'No, you don't.'

'I'm sure I do,' she said. 'Oh, it's moving.'

The newt was climbing through the branches of the twig, using its hands for leverage. It came to settle with its nose just breaking the surface of the water. The halves of its bronze eye glittered, separated by the horizontal bar.

'They're wonderful,' Isobel said. 'Little spirits of the river.'

I explained that they were found in ponds and streams, not rivers, but her words made me proud, and we went into breakfast chatting about the nets I had used, how hard it was to tell Palmate females from Smooth, and the landscape I would create for the newts. She seemed interested. Throughout that day, on the beach and in the car, I returned to a daydream. I lost it when we were swimming; when my father ran splashing into the sea towards us; but it came back when I was chilly in a towel in a deckchair, and again when I was walking to the shops. All that time, in my head, I was elaborating a story, in which Isobel and I helped the Narnians defeat an invasion led by *Fliegerleutnant* Kessel, whose Messerschmitts had got in through a portal in the sky.

The Palmate Newt is the smallest of the British reptiles and amphibians – rarely more than two inches long. Its name comes from those webbed back feet. To me, this seems an unassertive newt, not flashy like the others with their rippling crests and spotted bellies. This creature of the dark mud and submerged grassroots seems more subtle in its behaviour. Certainly its colours are subtle. I imagine the Palmate as a reserved creature, content to be inconspicuous; happy to let others take the limelight. From the scientific viewpoint, this is nonsense, of course. As far as we know, the Palmate gets as worked-up about mating and pursuing food as any other newt. But I am influenced by the fact that this is the newt of bare and lonely landscapes, the one you are likely to see in upland ponds and streams. And there you are only likely to see a few. I think of the Palmate as being at home in those windy, empty places, with their savoury drab colours – away from the crowds of other newts.

But it is not restricted to these landscapes. The Palmate can be found in most parts of Britain, though in East Anglia and the East Midlands it is rare. But on moors and heaths it is often the only newt, and in Wales, the West of England and Scotland it is much more common than the other two species. I was right to say that the Palmate was the newt of Dartmoor. The most likely reason for this distribution is that moorland usually has acid peaty soils, poor in nutrients. Ponds and streams there are more acid, and have fewer of the planktonic invertebrates – the water worms and 'water fleas' – that newt tadpoles eat. Palmate eggs and larvae are better able to thrive in these conditions than those of the Smooth and Great Crested – only slightly better, but enough, it seems, to make the difference. Apart from this, Palmate and Smooth Newts have the same ecological needs, and some lowland ponds, including garden ponds, contain both species. Occasionally, if the pond is wide and deep enough, there will be Great Cresteds too. When I've seen all three together, the Palmate has seemed the discreet one, slipping quietly into the party.

The old book from the farm bookcase gave *Molge palmatus* as the Palmate Newt's scientific name. In town that day, my father bought me *The Observer's Book of Pond Life*, which gave *Triturus helveticus*. All the books I found over the next few years used this latter name, which seemed to be the one modern experts had agreed on. But the name was an old one. Sometimes it was followed in brackets by 'Razoumoski 1788', which conjured up a lovely idea of early naturalists surrounded by abundant nature, theirs to name and classify. I wished it was still like that. The Palmate Newt was discovered as a separate species by Count Grigory Kirillovich Razoumoski or Razumovsky (1759–1837), whose splendid life story gives us a snapshot of that time in the European Enlightenment when aristocrats and statesmen wanted to be scientists, poets and philosophers also, seeing those pursuits as part of their role as leaders of humanity. Razoumoski was the fifth son of the last *Hetman* or elected Head of the independent Ukrainian Cossack state that broke away from Poland in 1649 and came under Russian protection, lasting as a semi-autonomous state until 1764, when Catherine II absorbed it fully into Russia. Deprived of his Russian citizenship as a punishment for criticising czarist rule, Grigory went into exile in Europe.

He noticed the newt in a pond in the Vaud, the French-speaking region of Switzerland; hence the word 'helveticus', meaning Swiss. This too is a lovely idea – that someone disappointed in their hopes of fame, power and great responsibility might discover the infinity of life in a small muddy ditch. It is one of our oldest stories: the greatness of the world can be found in the smallest of places. Razoumoski added 'helveticus' to the genus *Triturus*, which in Carl Linnaeus's taxonomic system was a classification within the *Salamandridae* family, which in turn belonged to the order *Caudata*. The name Triturus comes from the Greek god Triton, the son of Poseidon, the ruling sea-god, and Amphitrite, the sea-goddess and mother of seals and dolphins. Triton was Poseidon's messenger and herald, blowing a conch-shell for a trumpet. He was a merman, with a man's upper body and the tail of a fish, and he guided the Argonauts through treacherous marshes to the sea. Later myths replace the single character Triton with the Tritons, a species of men and women, gilled and fish-tailed, who form escorts for the sea-gods. It is touching that eighteenth-century naturalists, making the transition from enchanted nature to scientific nature, saw versions of these creatures in small newts in homely ponds. Isobel, seeing my newts as river spirits, was following this tradition.

Triturus means Triton-tailed, but for a time the name Triton itself was an alternative. Thomas Bell's *History of British Reptiles*, the first scientific guide to the British species, published in 1839, calls the Great Crested Newt *Triton cristatus*. Bell calls the Palmate Newt (or, for him, the Palmated Smooth-Newt) *Lissotriton palmipes*, and the names have come full-circle, for after *Triturus helveticus* had ruled for most of a century, three scientists overturned this rule. Garcia-Paris, Montori and Herrero proposed in 2004 that the genus *Tristatus* should be broken up, since the smoother newts were genetically closer to some other genera than to the warty *Triturus* newts with which they had been classed. *Lissotriton* should return, they suggested. The Palmate Newt would become *Lissotriton helveticus*, and the Smooth Newt *Lissotriton vulgaris*. *Lisso* is the Greek word for 'smooth'. The argument was persuasive, and the *Lissotriton* names have become the accepted ones and are in all the new literature. Scientific classifications are shifting sands. But *Triton* is still there.

About newts there are some strange traditional stories. Edward

Topsell, whose *Historie of Serpentes*, published in 1608, is a compilation of strange reptile-lore from ancient sources, says of the newt that:

Being moved to anger, it standeth upon the hinder-legs, and looketh directly in the face of him that hath stirred it, & so continueth til all the body be white, through a kind of white humor or poyson, that it swelleth outward, to harm (if it were possible) the person that did provoke it.

How could this tiny creature have inspired such a story? The Victorian writer of popular natural history, the Reverend J. G. Wood, relates in *Common Objects of the Country*, published in 1858, that English farmers used to swear that newts had bitten and poisoned their cattle. According to one story, a little girl filling a pitcher at a stream was bitten by an 'effet' that leapt from the water. The creature then spat fire into her wound. Her arm 'instantly swelled to the shoulder, and the doctor was obliged to cut it off'. It is easier to see how these later stories might have arisen. Before antibiotics, an infection from foul water could be deadly.

For my Palmates, the story ended badly. Isobel gave me a wave in the lobby, and said she hoped the newts would be all right. We drove to London with the can on the car floor, sloshing. Somehow they survived. I carried the can through the house, to the conservatory, and next morning set about building their new home.

What I wanted was a fish tank, but for the time being all I had was a tall glass jar from the sweetshop, the kind they still use for boiled sweets. It was smaller than the can, but would give me a side view of the newts. That jar had rather a grim history. I had used it before.

At the bottom of our garden was a park, and along the far side of the park ran a stream, called Chaffinch Brook. It was quite a fast stream, eddying in the middle, but in places, near the bank, there were pools that were almost still. This was the first wild place I had explored, the place where I first saw wild creatures and wanted to keep them where I could watch them. The summer before I caught the newts, I went fishing there for three-spined sticklebacks. I liked wading with my net, feeling the press of the water on my welling-tons. Some of the banks were shelves over the water, formed by tangles of tree roots. In the forks of these roots I would lodge my jamjar full of water. There it waited for my catch.

Sticklebacks could be seen in large numbers from the banks and the bridges. They foraged in the weed, making little darts and curving themselves around to nibble at it with tiny mouths. Their side fins fluttered. Silver sides flashed in the water. Supposedly, the males were building nests and tunnels in the dark filaments of weed, though I never saw this; it was one of the things I wanted to see behind glass at home. Redthroats, the older boys called them, because the males in spring and summer would flush bright red around the jaws. At the same time, their big round eyes became iridescent blue. In the most spectacular males, the area of the jaw became so red that it looked as if all the fish's blood had concentrated there, leaving the remainder of the body white and drained. A big redthroat was what I wanted, making a tunnel of weed at the bottom of the gobstopper jar I had prepared.

Lots of boys gathered to fish there, full of stories of what they had seen when venturing into the deeper parts, where the river flowed over my boot rims and I had to retreat and wring out my socks on the bank. The stream left the park by flowing under a roadbridge. We followed it, wading under the road, keeping close to the damp walls and hearing cars rumble overhead. On the other side of the bridge, the stream went on, into private land fenced off from the road with wire netting. The banks there were deep with nettles, and straggly trees trailing into the water. We couldn't see any shallows. To go under the bridge to the other side and look out at this wildness was daring enough. Bigger boys waded further, disappearing round the bend of the river. They said they'd seen pike there, grass snakes, otters, and a wild dog with mad eyes and a low growl.

The really big redthroats were out there, in the wild water, or occasionally in the dark water under the bridge. I usually hung back and heard other boys say, 'I can see one. It's a fuckun big 'un.' It was always a 'fuckun big 'un'. I didn't know what the word meant, and for a time assumed it was a word reserved for particularly spectacular redthroats, a word used by experienced fishermen, almost a technical term. At home I would tell my sisters about my exploits, and one evening, in the garden before teatime, I was saying to them, 'It was in the deep water under the bridge. I couldn't get to it, a fuckun big 'un,' and my father, appearing, said 'What? What did you say?' His face had gone twisted and white.

He made me repeat it.

'Go to your room.'

'Why?'

'NOW.'

My mother came up soon after. She questioned me about what the word meant. I said I didn't know, go away. Eventually, ashamed, I admitted it was something the other boys said when they saw a big fish. I didn't want to admit I didn't know. That's when I started crying.

She went down. I imagined my parents talking seriously down there. They came up together.

'Rich,' said my dad. 'You're a well brought-up boy. We want you to behave like one.'

'That word,' said my mum. 'It's not a word nice people use.'

'It's a word they used in the army,' said Dad. 'A violent, frightening sort of word. I heard it a lot in the war. I don't like hearing it. Not from you.'

'I didn't know.'

'I know.'

'Your teachers would be very shocked,' said Mum.

My parents' reactions were sometimes mysterious – exasperating to me, but mysterious in a way I couldn't question. About three years after the 'fuckun' incident, I was in the park with my friend Adrian. We were walking home after playing near the stream. A man approached us. Or was he a boy? He seemed something in between, seventeen or eighteen, perhaps older. 'Hello,' he said. 'What were you looking for?' 'Grass snakes,' we said, because although we knew there weren't any there, we often felt there might be; the banks looked so wild. 'Hmm, yeah,' he said. 'I saw a big one last week, over by the wall. I nearly caught it. Do you want me to show you?'

We went with him. There was nothing there, in the tunnel between the brambles and the fence. 'Do you like the Beatles?' he said.

We did.

'My dad works for the Beatles,' he said. 'He drives cars for them. Have you seen pictures of their cars? Do you know what they've got?'

'A Jaguar Mark Ten?'

'Yeah, that's right. Well, a Mark Ten for important trips. A couple of Mark Twos for ordinary. Have you got their LPs?'

We had some of them.

'I bet you haven't got signed ones.' This was true. 'I could get signed ones for you. If you'd like that. Would you like that?'

Yes, we would.

'Yeah. OK' – he seemed to make up his mind – 'OK. I can meet you here tomorrow at six. Can you do that? Yeah,' he said, when we nodded. 'That should be OK. I can get you all the LPs. Would you like to meet the Beatles?'

We thought we would.

'Yeah, that *could* work. But I'll have to test your muscles. We always have to do that. Look, there's a toilet just there. We could do it now, in the toilets. Won't take long.'

Somehow, I don't know how, I was expecting something like this. We looked at each other.

'Do we have to?'

'Well, yeah. It's the rules.'

'We can't tonight. We've got to get home.'

'Oh well.'

'You said you'd bring the LPs tomorrow.'

'Can't do it without the test. Well . . . All right. Six tomorrow. Meet me here. But you'll have to have the test.'

We walked off, leaving him standing there, and when we told my parents, we half-pretended we were telling them only because of our excitement about the LPs and meeting the Beatles.

An hour later, we were at the kitchen table with Ade's parents and a woman from the police. 'Did he touch you?' she asked.

'No, but he said he needs to test our muscles. He needs to, or we won't get the LPs.'

We looked from face to face, at Ade's parents, and mine, and the policewoman, and it was clear they wouldn't let us have our muscles tested. I realised I felt relieved, and looked at Ade. He shrugged.

What they wanted was for us to go to the place at six, like he'd said, but with the police watching from the trees and hiding at the park exits. This was exciting, but didn't seem real. Actually doing it was hard to imagine. What was going to happen? We were not to tell anyone at school.

Next morning, my father said, 'Oh. You've got Cubs tonight. I'd forgotten.' These were the Wolf Cubs, the junior Scouts. I went every week.

'But what about the police?'

'It's all right. I've told them.'

'I don't want to miss it, in the park.'

'It's all right. Adrian will do it. His parents don't mind.'

'Can't I miss Cubs, Dad? Just this once? It's important.'

'No. It's all arranged.'

'But Dad . . .'

'It's all arranged. You can't let people down. They're expecting you at Cubs. When you've made an arrangement, you keep it.'

So I didn't go.

'What happened?' I asked Adrian the following evening.

'He didn't turn up. The police hid in the bushes, and then I had to walk into the park on my own, and up to the bridge, and wait there. But he never turned up.'

We didn't talk about it. For some reason, we didn't want to. A few weeks later we saw the man again. The park was quite busy, that time. A crowd of boys were playing football.

He was standing near the football match.

'It's him,' Ade said. The man turned towards us, with a sort of recognition.

'You didn't come.'

'Sorry,' he said. 'I couldn't.'

'Our parents thought there was something wrong with it.'

'You told your parents?' He was agitated now.

'The police want to talk to you.'

'The police!' His face swivelled, left, right and behind. And he just turned, and ran away across the grass. We watched him get smaller. Afterwards, the only feeling I could identify clearly was anger with my parents for making me miss the stake-out. They were always stopping me from doing things.

After several of those stickleback attempts, I had caught the big redthroat I wanted. Mouth and cheeks were a deep orange-red, which was darker under the jaw. The body was silver green. In the jamjar, the fish raised its spines at me, gulping hard with its sad mouth. With him I brought home four females with swollen white bellies. Each cheek was a round silver scale-plate, protecting the gills and reflecting the light. I also brought a fistful of blanket weed, which billowed out, almost filling the sweetshop jar. Taking no notice of the ants' eggs I sprinkled on the water, the sticklebacks huddled under the weed, mouths opening and closing.

In the morning I hurried down, and saw at once that they were dead. Four were on top of the water; one tangled in the weed. Their mouths were gaping circles, hard and stretched. Spines on their backs were up, and below the gills, on either side, sharp bones stuck out like barbs. Round eyes were staring.

These were creatures of fast-flowing water. They needed aeration to breathe. I didn't know.

Already the jar gave off a rotting smell, mixed with the scent of pond water.

With the newts, I really wanted to be careful. It was lucky they hadn't drowned in the slapping water in the car. I knew that they would change after the breeding season, losing their breeding colours and climbing out of the water to live on land. There was no knowing when this would happen. So I placed a rock from our rockery into the jar, and poured in only so much water, leaving a third of the rock exposed. Emerging from the water, it was like the rock in the polar bear enclosure at London Zoo. At the bottom of the jar was aquarium gravel. Blanket weed hung in the water. I dropped in small earthworms for food.

Tomorrow my friends would come and see.

They came, and what they saw – what *I* saw when I ran down in the morning – were two very different creatures from the gold ones in the stream. On the rock were two yellow-brown lizard-like beasts, of rubbery texture. Gone were the green markings, gold stripes and high crinkly tails with threads at the tip. Here was a small neat animal, all one colour – though a beautiful delicate colour, with very faint marbling when I looked closer. The two newts flinched down, and kept still.

Overnight, they had changed from aquatic to terrestrial form. Their breeding season was over. I was taken aback by the suddenness. Had the journey shocked their systems? Perhaps I was right in this idea. These dramatic changes of colour, body-shape and texture happen because hormones come and go. Floods of prolactin bring the aquatic, sexually active phase, but when these levels fall, the body becomes more sensitive to thyroxin, produced by the thyroid, and the body-change begins. The different stages of metamorphosis in tadpoles – the appearance of limbs and the shrinking away of gills – come because of similar hormone changes. In adults and tadpoles, the hormones are responsive to nourishment, temperature, chemical signals in water and air, and seasonal cycles including hibernation.

I missed my golden flashing creatures, but I also liked the change. The two newts on the rock were more visible, more like animals at the zoo. They were living miniature dinosaurs, and once I had a tank I could give them a landscape of ferns and rocky outcrops. What would they eat, though? The earthworms now seemed too large, and those small brown lips too delicate. One of the books had said raw meat, so I dropped a red sliver on the rock. There was no reaction. The meat was there untouched when I went to bed. Tiny insects had to be the answer; greenfly perhaps. I needed to get the tank soon. But that didn't happen.

In the morning the jar was empty. The newts had got out. I fished out the weed, and then the stone.

Where were they? The jar was on a work-surface in the conservatory, from which a blue curtain hung down, hiding shelves of old paint tins, jars of nails and the like. An armchair was backed up against the curtain – a sun-faded old chair whose smell I liked. If I knelt in the chair, with my knees against the back and put my elbows on the top, my eyes were level with the bottom of the jar.

Now I lifted the chair away from the curtain. The floor was made of terracotta tiles. There was a straw carpet. I saw something, looked away and looked again. On a tile between the carpet and the curtain was the skeleton of a newt.

Tears pricked at my eyes. How could it be? What had happened? I was down on my knees beside the little huddle of bones, and found they were not bones, not exactly. The newt had shrivelled. It had become a dried-out corpse.

They have permeable skins. Under the microscope, their covering is more like netting and binding, holding their organs together, than skin. Newts have much less keratin than we do – the fibrous protein that makes up the skin's outer layer, and forms the scales on a reptile's body and human fingernails and toenails. Amphibians also have surface capillaries, tiny blood vessels whose linings are only a single cell layer thick. This means that oxygen from outside can pass into the blood through the skin, and carbon dioxide from the blood can be released. Amphibians breathe with their skins as well as their lungs. The skin is also permeable by water, which means that they need to remain in at least damp

41

surroundings, or their moisture will quickly be lost. This is what had happened.

I had made the most elementary mistake of assuming they wouldn't be able to climb sheer glass, and certainly not climb upside down as the glass curved into the narrow neck of the jar. So I hadn't covered the jar. I was wrong. When newts change into terrestrial form, their urge is to leave the water. Their bodies have considerable suction.

My fingers held the bony, rigid creature, and for a moment I had the idea that in water its flesh would fill out and come alive. Instead I put it into a matchbox, on a bed of cotton wool, like a jewel in a case, and buried it in the garden.

Chapter Two
Common Toad

———

My headlights turn the road yellow. Its surface is wet with spring rain. To the left, my light rushes along the base of the hedge. In front, insects veer and loop towards me and seem to be sucked under the car. Some seem to bounce on the road's surface, where objects catch the beam just before I am on them: stones, leaves and litter. A crumpled leaf is coming now, almost white in the light; no, it moves, it is a toad, and I am driving over it, hoping my wheels have missed it. Did I kill it? Here's another. And another. Is anything behind me?

I brake.

Hazard lights flashing, I step out into the damp dusty smell of the evening. My engine mutters. Where are the toads? Here's one, in front of the car, but lying twisted on its side, head crushed into the road. A hand sticks up, as if to fend off danger. The dead toad seems huge. I look away. But that one over there is alive, poised in mid-stride, a small female – the thin arms tell me that. When I loom over her, she shrinks, lowering her head and settling back onto her hind-legs, trying to be small or to dig herself into the ground. Toads do this, holding their heads down and shutting their eyes, with a wincing expression. *If I wait and don't look, then the trouble will pass.*

She is cool to my hand and heavy. Females on their way to the ponds are plump with eggs. Gulps pulse through her body. She is damp. I feel her squirt water on my hand, as toads do when picked up. People take this for urine, but it is water that the skin lets into the body and the toad stores in its bladder, since amphibians do not drink. Lifting her up, I catch her earthy scent and see the bronze eye, delicately veined, with the black oval pupil: the eye that traditionally made people think there was a jewel inside the toad's head. Sometimes they killed toads, looking for the jewel, expecting, perhaps, to find it in the red flesh, like a peach stone with shreds clinging to it. Of course, no precious stone could be found in this way, and various beliefs arose about the magic and ritual required for the extraction of the jewel. Edward Topsell, in his *Historie of Serpentes*, describes one recommended technique. The precious stone, he says, 'must be taken out of the head alive, before the toad be dead, with a piece of cloth the colour of red scarlet – wherewithal they are much delighted, so that while they stretch out themselves as it were in sport upon that cloth, they cast out the stone of their head, but instantly they sup it up again unless it be taken from them through some secret hole in the said cloth, whereby it falleth into a cistern or vessel of water, into which the Toad dareth not enter, by reason of the coldnesse of the water.' Topsell also records the belief that toads could grow and live inside the bodies of people who had swallowed toad eggs by mistake when drinking water.

I contemplate my toad in the wet yellow light. Under her jaw, the pale pouch fills and empties. Toads and frogs do not breathe with their chests. They take air in through the nose, and close their nostrils while using their throats to push the air back into their lungs. When the air comes back, the nostrils open, and the throat rises to push the air out. That is why the pouch under the jaw goes up and down. Her hind-legs push hard at my hand, trying to launch her body from my grip. On her back the tiny warts are almost prickly. She tries to climb out of my hand.

Headlights are coming up behind my car. My door is wide open in the middle of the road. I run back and slam it. The other car slows and stops. A young man gets out, his white tee-shirt seeming to float in the darkness. I wait in his beams.

'What's the trouble, mate?'

'It's OK. I'm rescuing toads on the road.'

'Oh. Can I see?'

I show him the toad. The pouch bobs again under her jaw. I have my finger and thumb under her arms now. Her sides bulge against them. To seem bigger to a predator, she has filled herself with air. 'Wicked,' says the man. He grins in the yellow light.

'We keep finding these in the pub toilets,' he says. 'There was one the other day, at my feet as I pissed. It almost fell in. I picked it up and took it outside. Loads get squashed here. I see 'em all the time. Uh, oh.'

Lights appear a long way off, coming towards us from the way our cars are facing. 'Better check that side of the road,' I say. The man follows. The beams of the approaching car pick out three or four corpses. 'Is that a live one?' says the man, pointing to the edge of the road. We both hold up our hands, and the car comes to a halt. I stride across. It's a male. I scoop him up and step back to the middle. We beckon the car on, and as it passes I hold up a toad in each hand in explanation. My friend gives them a thumbs-up sign. 'Can I hold it?' he asks.

'Yeah, here.'

He cups his hands, and I tip the male in.

'Aw,' he says. 'Aw. Whoops. He's peed on me.'

'It's water,' I say. 'They release it when frightened.'

'Aw,' he says. 'No need to be frightened, little mate.'

In the grass on the verge, we place the toads gently, side by side, facing into the bottom of a hedge, to them a forest. They don't move.

'Are you sure this is the right side?' says the man.

'Yes,' I say. 'Yes. Pretty sure.'

'They won't think, shit, we were nearly across and he's taken us back to the start?'

'No, I'm sure. They were facing that way.'

'OK. If you think so.'

He goes back to his car. 'Take it easy, mate,' he calls, and slams the door. Am I sure? Not completely. Perhaps they'll turn back to the road. I hope not.

Four more times it happens on this drive. Toads appear: dead ones, sometimes in piles, and live ones inching across. I begin to think every small rock or scrunched-up leaf in the road is a toad. Some probably are. Twice I see lots but can't stop; I'm upon them too quickly, with vehicles behind. How many I kill, I don't know. In one place, several corpses are tangled in a pile, limbs poking out stiffly. I rumble over it. Twice I am able to stop, leap out and quickly snatch up live toads, tipping them onto the verge before more cars appear. Tonight all the Common Toads in southern England seem to be on the move.

We call a wild animal 'charismatic' when it has intense meaning for us – traditional meaning, personal meaning, a mixture of the two. For me, Common Toads are charismatic, and their charisma is in their strange combination of homeliness and wildness. When I see one, I feel a pleasure of recognition, connected with childhood. They remind me of how I once felt that my home landscape, the safe landscape of England, was a place of excitingly wild depths, where, on a summer night as the sun went down, all manner of creatures might come out of the undergrowth.

A papery whirr overhead is a Stag Beetle in unsteady flight, its body hanging under the wings like a huge piece of machinery a helicopter is attempting to transport. Next morning, I will find the beetle on the pavement, black and shiny. Red antlers go up as my finger approaches. A rustle at the lawn's edge becomes a muttering hedgehog, stretching out as it snuffles left and right. Onto the garden table something drops, kicking and struggling: a quivering, velvety thing, with long feelers. It is an Elephant Hawk Moth, rose-pink, its wings like thick petals, its belly a twisting grub. And on the steps sits a drab little toad with pale gold eyes.

If an overgrown garden in the suburbs could produce such guardian spirits; if a piece of wasteland full of bramble and brick rubble could, then what was out there in the country beyond London? I imagined it: the thickets, forest glades, depths of bracken, over-grown rivers, marshes, moorlands, hidden valleys.

A toad turns its head carefully to look intently at a moving worm or insect. If the morsel seems desirable, a flick of the tongue carries the meal into the mouth, which shuts with a pop. Having no teeth, toads swallow their food whole. Woodlice, slugs, beetles, flies, moths and small earthworms go down in one gulp. Large worms are swal-lowed in several, the process sometimes taking take ten minutes or more; the toad rests between efforts, and the worm, hanging from its mouth, extends itself enquiringly and begins to slide out, until the toad swallows again. The toad cleans the worm by running its hands down the slippery body, the worm in between two fingers. With the worm all gulped in, the toad does a long shimmying stretch, eyes flattened, to finish the swallow. You can see the worm still writhing under the skin.

Common Toads have mild faces. They seem calm. Their outlines are rounded and gentle. Nothing is sharp. The little bumps that hold their eyes look touchingly vulnerable, especially from behind. Lost in deep concentration, a toad looks easy to take by surprise. Their saggy bodies are awkward, and they run jerkily or make short jumps that almost tip them over, but they have, none the less, an air of portly dignity, frowning comically over the efforts they have to make. Fussily, they settle themselves. With what looks like patient rumination, they sit watching the world, like elderly people in front gardens.

All this is anthropomorphism, of course. I am interpreting the toad's face and posture as if it were a strange kind of small human being. The meanings I find in the animal's features are meanings only human beings can perceive: a mixture of cultural meaning and meaning that is personal to me. These meanings belong to the world of my imagination and the world of human culture. They probably tell me nothing that is true about the toad, though they do make me give the toad a lot of attention, which leads me to learn more about it. That mouth and those eyes do not really indicate that the toad is patient and mild. They evolved through utility and accident. I know that I am not really seeing signs of how the toad feels – or only to the extent that we share basic forms of bodily response even with a small amphibian, such as shrinking from danger or rushing forward aroused by appetite.

How the toad sees *me*, I can barely imagine. I suppose it doesn't experience me as a single entity at all. It cannot see me whole. Toads have good eyesight; at night they can spot the movements of insects, slugs and worms two or three feet away. But, of course, their field of vision is close to the ground. Movement is what catches their attention: small movements, meaning food, and large movements, meaning danger. To the toad I must be an earthquake snatching it up, a wave of heat, or a hot dry confined space that descends around it. Lifted from the road by my hand, the toad might think it has been grabbed by a hawk's talons, except that the toad probably has no concept of 'hawk' either. Seized by a hawk, it must feel a strange weightlessness and a dreadful grip, piercing and slicing, but the toad cannot see the whole bird, just as on the road it cannot see the whole car; it has no angle from which to do so. Perhaps the toad never despairs. It does not see enough to know its plight. In the buzzard's nest, where it has been brought as food for the chicks, the toad wanders about, eats flies and settles down.

With human faces, I am naturally more cautious about taking set of jaw or shape of eye as evidence of personality. Yet it is difficult not to use ideas of what certain kinds of face mean: not to see a type of mouth as gentle or thoughtful, a type of jaw as resolute. Interpreting faces is an indispensable part of living socially, however cautious we must be. The language of facial and bodily gesture is essential to us as we relearn, all the time, who we are and who others are. One of the first things a child learns in the process of becoming social is the meaning of smiles and frowns, and those meanings are simple and crude before they can become subtle. Winston Churchill's face became one archetype: an icon of fearless, humane stubbornness. His face is often described as like a bulldog's, but it was also a rather toadlike face. From me this is not an insult.

There is a strange moment in Kenneth Grahame's *The Wind in the Willows*, the novel that features the most famous anthropomorphic toad. At the beginning of Chapter Ten, Mr Toad has escaped from prison, disguised as a washerwoman. Wary of pursuit after a narrow escape, he is walking beside a canal, hoping to find his way back to Toad Hall. A barge-woman, thinking he is an old washerwoman in distress, offers him a lift, and when he claims to enjoy washing clothes, she asks him to do a batch of washing in return. But he doesn't know how, and gets upset when the soapy water makes his paws crinkly. The barge-woman accuses him of fraud. Haughtily, he reveals himself as 'a very well known, respected, distinguished Toad':

> The woman moved nearer to him and peered under his bonnet keenly and closely. 'Why, so you are!' she cried. 'Well, I never! A horrid, nasty, crawly Toad! And in my nice clean barge, too! Now that is a thing that I will not have.'

And she throws him into the canal. I remember her voice on the LP of the book that we had when I was small. The voice echoes now in my head. Until this moment, the convention of the novel has been that the animals behave like human beings, and the people treat them as other people. The animals wear clothes, showing us the different social classes to which they belong. Animals and people have the same language and abilities, and share the same social world – though the boundaries seem to shift, and in some chapters animal society, though still anthropomorphic, seems more separate, and sometimes the animals are of human size and sometimes smaller. Still, Toad can own a country house, eat and drink with people in a pub, drive a motor car and be tried in court. But at this moment on the barge – and only for a moment – the distinction between human being and toad is back in full force, and the barge-woman sees Toad as a real person might see a real toad: that is, with a more normal degree of anthropomorphism, finding him 'horrid'. For just a second, *The Wind in the Willows* is a realistic novel.

Anthropomorphism can be flexible like that: a matter of constantly shifting boundaries. The anthropomorphic response to animals and the realist approach can lead to each other, frame each other. This is just as well, for we cannot do without either. Both are necessary parts of our response to nonhuman creatures, our way of living among them. For a moment there is a real toad in *The Wind in the Willows*, being viewed with unrealistic horror, as toads frequently are in real life. Whatever else this novel may be, it is a joyful response to a landscape that has in it real toads, moles, voles and badgers: animals that have entered the writer's imagination. Without the real animals there would be no novel. The poet Marianne Moore wrote that poems should present us with 'imaginary gardens with real toads in them'. A poem needs the free lyrical imagination *and* the intrusion of earthy, baggy, carbuncled, worm-eating reality. Moore's reality principle is represented by a toad.

That is the puzzle of anthropomorphism. We need it and we need to overcome it. A dolphin's face seems to wear a friendly smile. We know that this smile is not really a smile, but a shape of face and mouth that evolved for other reasons. The line of the mouth does not really mean that the dolphin is friendly. We need to acknowledge this. But the idea of suppressing our recognition of the mouth-line as a smile, to the point of banishing that recognition altogether, seems a violent idea to me. The recognition is rooted in our earliest learning of how to make sense of the world, how to live there. We need to see the dolphin's expression as a smile *and* to know that it is not.

One of the cultural functions of wild animals for human beings has always been to provide archetypes of different sorts of moral character – the cunning predator, the foolish predator, the worker, the idler, the boaster, the self-sacrificing mother. Such figures make up an allegorical landscape of clear moral categories. They are land-marks that tell us where we are. If that moral landscape is not continuous with the real, material landscape in which we live, we are estranged from that real landscape: not really engaged with it, but living in a separate mental world. With wild animals there is, traditionally, a licence to use stereotypes, since – usually – the animals have no power over us, and we are not wondering whether they will become our personal friends or enemies. I interpret the toad's face – its generic features more than those of any particular toad – in anthropomorphic ways because I have only a weak sense that toads have individual personalities.

How, then, are we to respond to animals in human terms – emotional and imaginative terms, as well as scientific terms – while also seeing how different from us they are? Anthropomorphism has its dangers, of course; not least to the animals themselves. I find toads likeable, but there is a long tradition of associating them with evil. Witches were believed to keep them as familiars. Sometimes toads were the Devil in disguise. He appeared in toad form to be kissed on the lips by the witches. Toads had a Gothic, gargoyle-like quality. To the popular medieval Catholic mind and the later Puritan mind also, the world was teeming with small demonic beings, peering maliciously from the undergrowth. The migrations of toads through the fields must have seemed like the movements of goblin armies.

Thomas Pennant, the eighteenth-century naturalist, described the toad as 'The most deformed and hideous of all animals; the body broad, the back flat, and covered with a pimply dusky hide; the belly large, swagging, and swelling out; the legs short; and its pace laboured and crawling; its retreat gloomy and filthy; in short, its general appearance is such as to strike one with disgust and horror.' Pennant was rebuked for these words by Thomas Bell, in his wonderful *History of British Reptiles*. But Bell himself, determined to find in all wild creatures the evidence of a beneficent Creation, can only call the Common Toad 'far from prepossessing'. 'The body,' he says, 'is puffed out and swollen; the head large, flat on the top; the muzzle rounded, and very obtuse. There are no teeth either in the jaw bones or on the palate. There is above the eyes a slight protuberance, studded with pores; and the parotids' – these are the poison-glands behind the eyes – 'are large, thick, prominent, and porous, secreting an acrid fluid.' Bell is attempting to be scientifically dispassionate, but words like 'dirty' keep turning up in his description. His attitude, underneath, is a bit like that of the barge-woman. Bell says that the toad's 'upper parts are of a dirty, lurid, or blackish grey, with sometimes a slight greenish tinge; tubercles' – these are the warts – 'more or less brown; beneath dirty yellowish white, sometimes spotted with black.'

Shakespeare's Othello, deceived into suspecting that his wife Desdemona has taken a lover, feels that his centre, the shyest part of him, his most sacred emotional place, has been defiled. An intruder has dirtied it.

But there, where I have garner'd up my heart,
Where either I must live, or bear no life;
The fountain from the which my current runs,
Or else dries up; to be discarded thence!
Or keep it as a cistern for foul toads
To knot and gender in!

When I first read these lines at school, I laughed aloud. I knew those foul toads. They mattered to me. I saw the image; I knew what this 'knot' was. Male toads congregate in the ponds in spring, squeaking rapidly, heads just above water. Each is desperate to fix himself to a female's back. At this time of year, they certainly are not calm. Hormones are surging, pheromones tugging. Five or six males, squeaking in crescendo, converge on a female, grabbing at her from all sides. If a male is already on her, they try to force their heads in under him, to prise him off. He kicks at them, and tries to steer the female away, but three or four rival males have already taken hold. The female is overweighted now, and the knot of toads rolls over like a ball. More males arrive, as many as ten, clasping on where they can, arching their backs and squeezing tighter. They grapple and strain. The water boils with their struggle. She is buried at the centre. Often she drowns. They squeeze the breath out of her, and stop her from surfacing for air. Males in these frenzies grab anything. Sometimes frogs are pulled in, even fish.

Once I tried to save the female. I was in my thirties, teaching at a school in South London, and had taken a small group of students, five or six of them, aged fourteen or fifteen, for a natural history ramble on a common that was popular with dog-walkers: a bit of gorsey heath, a bit of woodland and a golf course. Walks like this were one of the things we did on the afternoon reserved each week for General Studies. There was a pond at the bottom of a grassy crater, and we scrambled and slid down the bank to see what life there might be among the brown reeds. Immediately I heard male toads squeaking, and pointed out a small group of them, sitting on the pondweed. My students, hugging themselves in their coats, seemed unimpressed.

'I've got mud on my trainers,' said one. 'And my tights! Aow. My mum'll kill me, sir.'

'What's that over there?' said another.

To one side of the pond, in deep water and dark weed, with the sandy bottom looking slightly bluish, was a rolling ball of toads.

'What are they doing?'

'Yeah, what are they doing, *sir*?' That was one of the boys.

I explained. A burst of squeaking came from the toads, and the ball rolled towards us.

'Oh God, sir. Can't we do something?'

'I think you ought to make them stop, sir.'

'Yeah, sir.'

'All right.' I grinned nervously. 'All right. I'll try.'

Leaning out across the pond, I used a stick to manoeuvre the ball towards me. It bobbed in the water. My hands were shocked by the cold, but I got them under the ball, and felt the softness of the struggling bodies. In response to my touch, the toads bound themselves tighter. I lifted the whole writhing mass from the water, and tumbled it onto the grass, in a tumult of squeaking. The ball had many voices. I expected it to come apart out of the water, but none of the males would let go.

My students had all gathered round.

'God. Look at them.'

A greenish male was on top of the ball, his glistening skin

quivering. In his yellow-gold eye, the black pupil had flattened into a line. It made him look sexually drowsy. His squeaks sounded tired and tinny, but went on. I tried to get my finger under his gripping arm, to break the tension of his hold, but his arm was quite rigid. He gripped harder, his eyes disappearing under his cheeks with the effort. I was afraid of snapping his arm.

'What are you doing?' At the top of the bank, a woman looked down at us. She looked about forty, and was wearing a long scarf. Crinkly blonde hair. Beside her a big dog, a lurcher, pulled at the lead. 'I don't believe it,' she said. 'You're trying to stop them. Leave them alone. It's natural.'

'No,' I said. 'No. I'm trying to save the female. There are too many males on her. They could drown her.'

'Really?' she said, suspiciously. 'Are you sure?'

'I've seen it happen.'

'It's true,' piped up one of the students.

'Oh,' she said. 'Well . . . Oh. Can you get them off her?'

'I don't think I can. They are gripping too tightly. I'm afraid of hurting them.'

'Better put them back, then,' she said firmly. She was right. I picked up the ball again. In my hands it struggled like a single creature. The students watched in silence. A toad's leg forced its way between my fingers. At least the female had probably taken some breaths. If I left the ball out of the water, crows would probably attack it. I placed it back into the pond.

Othello looks at the writhing knot with a horror that is different from mine. I fear for the female. He sees the toads as sexual appetite incarnate; nothing but sex animating their bodies. This pure brutal impulse terrifies and humiliates Othello – the absence of hesitation, tenderness, shyness, or concern for others' feelings. Toads behaving like this are a disgusting, defiling presence. In Shakespeare's time, the ancient belief still survived that these animals were spontaneously produced by decomposing marshy slime. Edward Topsell, in his *Historie of Serpentes*, says that, according to many witnesses, toads are 'bredde out of the putrefaction and corruption of the earth'. Toads were not only the evil familiars of witches and a form often assumed by the Devil. They were the living demonic essence of dirt and decay.

That element of Gothic eeriness, at the edge of my perception of toads, is part of their charisma for me. At the age of five, I was given a book called *The Wonder World of Nature*, a compilation of chapters on natural history topics by a long list of expert authors. It had been published in 1959, the year before. Even as late as the 1960s, wildlife books often had photographs that were badly focused and murky. In this book, the illustration of the Common Toad shows a squat black lump, of unclear shape, sitting on a bare stone surface. It looks like a pile of something put there with a spade. Just discernible at the front of the lump, if you look hard, are thick raw forearms and pointed fat fingers. Somehow, the darkness and poor focus have given the toad an evil eye and grim mouth. No other features are clear. The toad looks like a small wicked face on legs. I felt a thrilled shudder, to think these were lurking outside, in the garden, the park, and the country beyond.

This negative anthropomorphism that identifies certain wild creatures as demonic is the flip side of the foolishly optimistic and cuddly anthropomorphism – the kind that is so often called 'sentimental'. Routinely, that word is brandished against people who want to protect wildlife. They are charged with turning to sentimental anthropomorphism in order to avoid seeing the cruelty and indifference of nature. Yet the idea of nature as a force or personage that is collectively 'indifferent' is itself anthropomorphic. Nature, collectively, does not have an emotional or moral stance of any kind, and individual wild creatures are certainly not indifferent when they see us as danger or prey, or engage with us in any other way. What is generally meant by nature's 'indifference' is that animals do not empathise with us and care about our suffering, and doubtless that is nearly always true, but it does not make the animals indifferent. Wild creatures have myriad responses to human beings, and involvements with us. In the determination to see nature as indifferent, there is sometimes a reverse sentimentality: a self-congratulatory air of masculine toughness and an impatient dismissal of all considerations that stand in the way of commercial exploitation. The assumption often seems to be that if nature is indifferent, we should be indifferent in return. It is a convenient assumption.

Nature lovers should always think carefully about the accusation of sentimentality, but not be bullied by it. Often, the love of wild nature as dramatic spectacle is not an avoidance of mortality and random accident – that randomness is what is most seriously meant by 'indifference' – but the opposite, a desire to find a safe way of acknowledging these things. Wild nature, as a spectacle in which we are not directly engaged, offers this opportunity.

Another charge against anthropomorphism, a related one, is that the practice expresses a human arrogance – an assumption that we are the model of which all other species are deficient imitations. Anthropomorphism is a symptom of a preoccupation with ourselves that makes us uninterested in the otherness of other creatures. Again, it is a charge that nature lovers must always examine with care. Anthropomorphism can be taken so far that the meaning or personality given to the animal no longer represents any sort of attention to the real creature. Cartoons and heraldry do this. And for some animals traditionally cast as evil in the moral dramas that we make of the world, the consequences have been catastrophic – not for toads, much, since in practice they are no danger to us and come into no competition with human beings, but certainly for snakes and wolves, which in Europe and North America have been persecuted far beyond any actual threat they have posed.

But simply to renounce anthropomorphism – to attempt to stop doing it – would be to separate the natural world from the world of human meaning. It would be a refusal to recognise wild nature as part of our home, where we find emotional sustenance. 'Keep out,' we would be saying to one of our most fundamental appetites for meaning. 'You are not allowed to seek nourishment here.' Anthropomorphism is one of our most fundamental ways of engaging with the world. The trick is not to seek to expel it, but instead, when we use it, to give precise attention to each of those dangers, and keep our anthropomorphism in contact with our perception of the otherness of wild creatures.

Common Toads are not spontaneously generated by mud. There is, however, an earthiness about them. It makes sense to see them as distillations of earthy environments, for in those environments their colours have evolved. Like soil, they soften and swell in the rain, and tighten and go dusty and drab in dry weather. Their bodies are the colours of different kinds of earth. Lots of the males are a sandy-muddy colour, suggestive of yellowish clay. Some have an olive-green tinge. Females usually have more complex mixtures. Dark browns and light browns intermingle in streaks and patches: burnt brown, reddish brown, clay, fading to near-white at the sides, and all of it darkly freckled and splashed. The skin is a landscape, textured with colours. Large warts are often pale, as if the brown has trickled off and pooled around them. On the toad's white under-side, where the skin is a surface of tiny smooth bubbles, there are speckles of brown, dark as chocolate.

I find these toads comforting: unflustered, vulnerable spirits of the earth. George Orwell found them comforting too. In his essay 'Some Thoughts on the Common Toad', published in 1946, he praised the toad as a herald of the returning spring, amid the rubble and exhaustion of blitzed London. His toads are ordinary Londoners, carrying on. It is their ordinariness that touches him: the ordinariness of their knobbly drab appearance, and the ordinariness of their heroic unquenchable liveliness, continuing to thrive in corners and crannies, and emerging each year with the same undignified enthusiasm. They represent wild nature as a common possession: ordinary wildness, not fenced-off but public, shared by all. Orwell's essay is one of many works of natural history written at that time that have a democratic spirit. Britain's wild nature was seen as a shared inheritance for all who had survived the war. 'The point,' says Orwell, 'is that the pleasures of spring are available to everybody, and cost nothing.' This is still fundamentally true.

The first warm, wet, still nights get them moving, usually in early March. Warmth and moisture: something in the air. Dug into the earth, huddled under stones or logs, or using empty burrows, they wake up, and their bodies change.

Amphibians breathe partly through their nostrils and partly through their moist, porous skins, with which they take up oxygen and release carbon dioxide. Frogs can hibernate under water, breathing only with their skins. Amphibian skins also absorb and release moisture. This happens rapidly, which is why the animals can quickly dry out, despite their numerous mucus glands. Frogs are especially vulnerable and must stay in wet places, even though their mucus glands are the most productive, keeping them slick and sticky. Toads, having dryer skins, less open to the passage of water and gases, do not need so much moisture around them, but their surroundings must at least be damp. They very rarely winter under water, and, when active, they cannot breathe for long with their skins alone; that is why the female toad drowns in the ball. Toads and frogs look saggy because their skins are not attached all over their bodies, but only along lines that partition the inner surface. The skins of the males, especially, become loose and soft in spring. When a male toad or frog clings to a female, he quivers like a bag of liquid.

Spring arouses them and readies them for breeding. Bellies sag with eggs. Males now have thickened forearms and hard black bumps on the inner parts of their fingers – nuptial pads, the books call them. These are for holding on to the female. The arms will pass under hers, almost meeting, and the hardened pads grip deep. His back arches to fit hers, as he locks into position. If he gets it right, he is difficult to dislodge. His hind legs can paddle her through the water and kick rivals away.

But that is for later; that is for the ponds – though some lucky males mount their females on the way. First, something makes all the toads set out for the water, and guides them there. Not much is known about this. Scents from the pond reach their nostrils, no doubt, in the warm night air: tiny particles conveying the essence of slime, rotting pondweed, and pond plants beginning to grow. Experiments with frogs have found them to be attracted by small quantities in the air of glycolic acid, a chemical produced by algae in ponds. The frog or toad is drawn on, as the scent becomes stronger. Most likely, the sensation is much stronger than anything smell can give us. Perhaps the animal feels the chemical all over its skin, since amphibian skin is much more open to airborne and waterborne chemicals than ours. And perhaps the ponds have a magnetic pull, quite literally. It is possible that, as in migrating birds and fish, magnetite in the toad's body responds to the earth's magnetic field, so that the body recognises patterns and steers the animal in the right direction. Sounds play a part as well. The first males to reach the ponds start calling, though the male Common Toad does not send out a long resounding trill or cackle, like the Natterjack or the Marsh Frog. He makes short squeaking croaks, rapidly repeated – unimpressive in comparison, but endearing. In various ways, the pond summons its toads.

Whatever the mixture of information that stirs up their hormones, softens their skins and draws them towards the ponds, imagine what the call must be like for these toads. The pond is far away, perhaps almost a mile, but to the toad it must be an immense presence. They are pulled by the nostrils. It fills up their brains. They smell it; they hear it; why haven't they reached it? Its sound is a roar. It is music and sex. And it draws them onto the roads.

Not all return to their birthplaces. A few colonise new ponds, perhaps drawn from their track by the scents. Or a male heading for one pond may board a female going to another. Some males disappointed at their first pond go looking elsewhere. Most, though, return to the pond they left as toadlets. There they face the scrambling battle. Once it is over – because most males are paired, or a strong male has fended off the others – they are ready to spawn.

They swim into weed or underwater roots: something to wrap the strings of spawn around, so that the precious material will not drift or sink. When the female is ready – it must be when she feels the spawn coming – she warns the male by straightening out her back beneath him and stretching her legs. A male amphibian has no penis. In both sexes, waste matter and reproductive material, eggs or sperm, emerge from an opening between the legs called the cloaca, which becomes plump in spring. She straightens her back and stretches out her legs so as to bring her cloaca up under the male's cloaca. He responds by positioning his feet so that his toes touch her cloaca. Two jelly strings begin to emerge, full of spherical black eggs. The male's feet feel the jelly. It is that touch – the touch of jelly on his toes and the soles of his feet – that brings him to sexual climax. Sperm shoots from his cloaca onto the two emerging strings. She rests, perhaps goes up for air, and then squeezes out more. Whenever more jelly comes out, its touch makes him ejaculate again. He paddles his feet to distribute the sperm. The process takes several hours. With the spawn trailing, the pair of toads moves through the weed, so that the strings of eggs get tangled with the stalks. At last she is done.

The strings drop away. His feet feel no more jelly. He paddles for a while. Nothing comes. He releases his hold on the female.

The adults leave the water. In their jelly, the eggs change shape. The black dot becomes a kidney shape, one end of which grows into a head, the other into a tail. The tadpoles begin to twitch. They curl and uncurl, flipping around in circles. After about a fortnight, the tadpoles are ready to break free. Glands on their heads exude protein-dissolving enzymes that loosen the jelly, enabling the tadpoles to force their way out. Still embryonic in shape, they hang from the empty strings, their bodies slowly furling and unfurling.

A day or two later, they are swimming freely and have the familiar tadpole shape: from above, a black pea-sized body with a wiggling tail, from the side, a body shaped more like a guinea-pig. They are smooth and absolutely black. Their skins release toxins: *bufotoxins*, from *bufo*, the Latin for toad. In adults, these toxins come out of the wart-like bumps all over the toad – not really warts, but glands – especially the long grub-shaped ridge behind each eye, the parotid gland. The toad exudes the chemical if bitten or roughly seized. A cat or dog will recoil from the nasty taste and drop the toad, though many other predators, especially birds, attack without inhibition.

Newts and fish turn away, sensing the toxin in the water. If they do put mouth on a tadpole, the taste repels them. But insect predators are another matter. Many are able to pierce the body and suck out the contents, leaving the toxic skin draped on the weed. Diving beetles and their larvae do this, and water scorpions; dragon-fly larvae, too, firing parts of their mouths at the tadpoles on a retractable jaw, harpooning their meal. Often the tadpoles mass together: a soft, gently rippling black ball on the yellow clay of the bottom or among the light green plants. In large ponds the shoal can be several feet wide, a mass, basking, inert, near the surface or moving along, all wriggling in the same direction. Once I netted the mass. As soon as the tadpoles were out of the water, the weight nearly broke my net. It was like a trawler's net of fish. They poured into my bucket like oil.

Linnaeus named the Common Toad *Bufo bufo*. That is, the type of toad that is just called 'toad', the first toad, the plain toad, the simple archetype of a toad, the model of what a toad is, upon which others are variations. In Europe, *Bufo bufo* is the most numerous toad, the ordinary toad, found commonly from Britain and Iberia in the West (but not Ireland) all the way across Europe and far into Russia, as far as Lake Baikal in Siberia. They hibernate under the Arctic ice of Norway and Sweden, and hide from the sun in North Africa. Europe has three species in the *Bufonidae* family: *Bufo bufo*, *Bufo calamita*, the Natterjack, and *Bufo viridis*, the Green Toad, which has lovely distinct green blotches on its back. Some books call these *Bufonidae* Typical Toads, which pleased me as a child – it made them seem solid and homely, and that's how I already thought of the Common Toad. Other books call the *Bufonidae* True Toads, which implies, strangely, that the other European toads, the Spadefoots (*Pelobates*) and Fire-bellies (*Bombina*), are somehow not quite toads at all. This idea also adds to the earthiness and solidity of the Common Toad, *Bufo bufo*. In Spain and Portugal and along the Mediterranean coasts of Europe and North Africa, there is a subspecies, *Bufo bufo spinosus*, which grows larger and has tiny thornlike spines on its warts.

There are two toad memories I cannot place. One must be from quite early childhood, when I was five or six, perhaps. I dream it, sometimes. My family were all present – father, mother, sisters, and my father's parents, perhaps my mother's too, a crowd of us. We were crossing a field near a river, or having a picnic. I think it was near Flatford or Dedham in Suffolk. My father's parents lived in Ipswich, and until my grandmother died when I was seven, we often holidayed there. So I must have been younger than seven. In that field, we saw maybe fifty toads in ten minutes. To me it seemed like hundreds.

I half-remember, half-imagine it like this. 'Look, Rich,' my dad says, crouching down, pointing into a tussock. 'Here's something that would interest you. A toad.' I crouch down with him. His finger points out the knobbly dry skin, the bright eye. I can see his finger now, against the grass, and the back of his hand – the way the skin wrinkled and the thick veins stood out soft around the knuckles. Now the back of my hand looks like his.

'Oh, there's another one, look,' says my mum. And now we are all seeing them, everywhere we look. Toads are all over the place in this field. Every tussock seems to hold one. 'It's alive with them,' my mother gasps. 'Like a biblical plague,' says my dad. '*Toads*,' says my grandfather, sucking his pipe. 'Oh yes. We used to see lots of toads in the fields round here. I haven't seen this many for years, years. Teeming with 'em. Look at that.'

We did. There were toads of different colours: clay brown, speckled brown, dark green, speckled green, some almost black. As we stood over them, they hunkered down and puffed out their sides. They were running in front of our feet. 'Here's one!' I kept calling out in delight. I trotted a little way on. 'Here's another one.' A bit further. 'Here's one, a really big one!' And so on. I was desperate to take one home and keep it. 'Just one,' I pleaded. 'No, no,' said my mother. 'They live here in the field. It would be cruel to keep one in a jar. And to take it back to London, from its place here by the river.' At that age too young to find arguments, I accepted what she said. But when we reached the gate, I refused to leave the field. 'No,' I shouted. 'No. A bit longer! I want to see another one. No!'

My father was beginning to feel helpless. In his work as a school-teacher he struggled with this feeling, I now know. 'Come on, Richie,' he said. 'Come on now. You've seen enough toads. We've got to get home. Come on now.' His voice was getting tenser. 'Your grand-parents are getting tired. Come.'

I held on to the gate. 'No, no. Please. Just five more minutes. Please.' My words tumbled out in panic.

'Oh. Well . . .' He looked around. 'No. Look, no. Obey your father.'

'Five more minutes. Please!'

'No!' He jerked me roughly from the gate. The others were all watching from the car.

'Oh, it's not fair.' I was crying now. Twisting my hand round in his, I broke free, and ran back to the gate, determined to see one more toad, but he grabbed me around the waist this time and bundled me to the car.

'Just get in,' he said. He slammed the door.

We must have been in the path of the migration, and for some reason – perhaps moisture in the air, and sudden warmth – the toads were travelling by day. I have seen nothing like it since. Several times, in the years after that, I dragged my family to that field and found nothing. It's not the right one, I told myself. The fields all looked much the same. Similarities tricked me. Here was a stile that might be the one. That deep stand of nettles looked familiar; that bend in the river. But there were never any toads.

My other memory is more unpleasant. I really have no idea where this took place. It might have been in our London garden, or perhaps we were visiting someone. I am sure I was older than when I saw the field of toads, but it was before toads became familiar to me, before I kept them as pets and knew how to find them in the wild. Again, I was very excited to see one. This one was large. I have a memory of seeing it sitting on a lawn, and then of gazing at it in a large glass jar. Someone called me over to see it, and caught it for me, I think. And I noticed the toad's nose was a funny shape, rounder than it should have been, bulbous. I looked closer. The nostrils were too large. I could see into them. There were worms inside, tiny white worms, wriggling.

I had read about this. There was a fly that did this to toads. I looked it up later, and learned that it was called the Toad Fly – a name that spoke of a cruel world in which each creature had its allotted nemesis. And the fate of a toad attacked by this fly is exceptionally cruel. The Toad Fly is an iridescent green-bottle, *Lucilia bufonivora*, the toad-devourer, the fly that lives on toads. Touching down lightly on a toad's lower back, out of sight and out of range of the tongue, *Lucilia* attaches to the toad a cluster of long white eggs, like grass seeds. They resemble the eggs other blowflies leave in rotting meat. Maggots hatch out, and advance up the back and crawl into the nose, where they settle and begin eating their way through the nostril walls in all directions, reaching the brain after several days. The nostrils become open craters. It is a horror. Unable to breathe through its nose, the toad makes large gulps with its mouth. It runs about, no longer cautious. Perhaps there is relief in movement; perhaps stillness is intolerable. Blackened corpses are found with great holes in the nose or the top of the head, where the worms have exited. They look like shell-holes in the turrets of burnt-out tanks. *Lucilia* is quite uncommon in Britain, but studies of some toad colonies on the Continent have found as many as eight per cent of the toads to be carrying the worms, mainly the larger adults. I have never seen an affected toad since that one on a lawn when I was little. The memory became half-mythical to me.

Nowadays we do not consider some animals to be demonic, but if ever there was a candidate for rational demonisation, the Toad Fly is one, along with the plague bacillus, the Guinea worm and numerous viruses. Yes, the fly is merely reproducing in its evolved fashion. Yes, it has found an ecological niche, a livelihood. But what a truly dreadful niche it is. From the viewpoint of many animals, however – hens and pigs would be obvious examples – the species that ought to be demonised is the human.

I pestered my father into buying me a pond for the garden. We bought it in spring, one year after my newts had died. The pond was a mould of rigid grey plastic, one of the smallest they had, about four feet long and three across, with shallows around the edges dropping to a depth in the middle of two feet. I sank the pond into a grassy patch at the fork of two paths, near the old pear tree. Before filling the pond with water, I put in spadefuls of earth – I know now that this is the vital thing if you want an ecosystem with oxygenating plants and good visibility. In a couple of weeks the water cleared, and I added pondweeds anchored by stones, pond snails, three goldfish and a bucket of newly hatched frog tadpoles; a neighbour had invited me to take some from his much larger pond. In the ensuing weeks, the fish ate nearly all the tadpoles. I had put in far too many. Regretfully, I watched the fish suck tadpoles in by the mouthful, chew once or twice, blow them out in a swirl of broken-backed bodies and suck them back in. When the fish disappeared, presumably clawed out by cats, I was not very sorry. I saw them as a beginner's mistake.

Gradually, the pond came to life. *Daphnia* water-fleas appeared in shafts of sunlight in the water, clouds of them. They swam in little jumps. Mosquito larvae hung from the surface, doubling-up and straightening. The few surviving tadpoles wiggled peacefully here and there, growing fatter. One day, to my amazement, only a few weeks after starting the pond, I saw a newt in the water, a small dun-coloured newt that I identified as a female Smooth Newt. Where had it come from? In the mornings I ran down to gaze into the water.

One evening each week, I went to the Wolf Cubs at a church hall down the road. Another boy from my school was there, Brian Myerscough (pronounced 'myersco'), a dark-haired boy with cheeks that turned ruddy when he was out of breath. Microscope, everybody called him. 'It's fucking Microscope,' the bigger boys would say. To me, he seemed quite a tough boy at first, but there was an anxiety about him, an eagerness to please. He invited me to his house after school one day. We had baked beans on toast – white sliced bread from a bag – with a thin square slice of cheese on top, and ketchup. The cheese slices were Kraft Cheese Singles, mustard-yellow. I looked at the packet with fascination. Each slice was separately wrapped in plastic. It was cellophane, clear as glass. You took hold of the corner and gently peeled back. The wrapping came away with a faint smudge marking its surface. Next morning, at home, I asked if we could get some Kraft Cheese Singles, and my father overheard. His face tensed.

'Kraft Cheese Slices,' he said. 'We're not having that rubbish in this house.'

'It's Kraft Cheese *Singles*.'

'Whatever they call it, it's mass-produced junk, not real cheese.' My father would not have food like this at home. 'Processed cheese,' he would snarl. 'Sliced bread! It's bloody cotton wool.' These things made him angry. He feared them. I was beginning to understand why.

We played with Microscope's collection of soldiers and weaponry, which impressed me; a much better collection than mine. His garden was less impressive: small, mainly concrete and gravel. At the bottom, beyond a high wire-netting fence, was the railway bank, covered in brambles. It looked wild. 'Is there a way to get in there?' I asked him. 'There might be animals.'

'Yeah, there are foxes. We hear them at night. It's dangerous,' he said. 'Big kids in there – they dare you to stand on the track until the train's nearly there. They hold you.'

As we went back to his kitchen door, something through a window caught my eye. I went closer. It was a room I hadn't been inside. On the table were rows of toy animals, the small plastic wild animals I loved in the toyshops. In long rows like soldiers there were lions, each with a snarling face and one paw lifted. Behind them, rows of

elephants, rhinos, hippos with wide pink mouths. I was delighted.
'Look at those!'

'Oh. Yeah', said Microscope.

'Can we play with them?' I had never been able to assemble herds like this, and had visions of them in our garden at home, around the rockery – a Serengeti.

'No. They're my granddad's. He paints them.'

On a trolley were little pots of paint, rags and brushes. I saw that the lions had red mouths, black lips, black eyes and brown manes only halfway down the row. Beyond that they were blank-faced yellow plastic.

'Come on,' said Microscope. 'He doesn't like people looking.'

A week later, Microscope came to our house for tea. My mother cooked liver, bacon and mashed potato, with cabbage. We went down the garden and crouched over the pond. 'Wow,' he said. 'Wow.'

Pleased, I ran my hand around the rim, lifting the grass that already hung thickly over the edge. 'This is where you sometimes get frogs and toads,' I went on. 'Under this grass.' I didn't know I was right.

Something moved under my hand, cool skin, a trembling body in the white grassroots. My hand came out, holding a small toad, about an inch long. It was beautiful: a lovely light brown. Its red-golden eye shone. On its belly the pattern was a mosaic in exquisite miniature, made of thousands of little bumps too small to see. Covering the toad with my hand, I carried it in triumph to the lawn. I had found it in my garden. Toads like this were living there.

'Let me see, let me see,' said Microscope. 'Wow.'

'This is one of last year's hatch,' I said. 'It's in pretty good condition. I am hoping that this one will breed next year. It should be ready.'

'Wow. Can I hold him?'

'Just for a minute. Then we'd better put him back.'

I was ecstatic to have found this toad, which I had never seen before. It was wild. Our garden was wild. In bed that night, I thought of the toad, out there, moving about, hunting for insects, then returning to the roots at the pond's edge. But next morning, the toad wasn't there. Somewhere else in the garden, I thought.

At school, in the afternoon, Microscope took me aside. 'I've got something to tell you,' he said. 'Last night, after my mum collected me, we went for a walk in the park at the back of your house. Somehow I knew that something would be wrong. I was right. Someone was climbing out of your garden. It was a big kid, a teenage kid, from Allen Park.' This was a school nearby. 'He came jumping down from your fence, with something in his hand. I asked him what it was, and he showed me. It was your toad, the one we found.'

'What?'

'Yeah. But don't worry. I got it off him. I told him you were my friend, and I said I would go to the police. He gave it to me.'

'Where is it?'

'Round at ours. You can come tomorrow and get it.'

'Today,' I said. 'Straight from school.'

At his house, the little toad was in his bedroom, under a handful of grass in a large glass jar.

'What did your mum say?' I asked.

'Oh, she didn't see. She was having a smoke. I ran off on my own.'

'Thanks,' I said. 'Thanks.'

'It was good I was there.'

'Yes, it was.'

'Could we get some?' he said. 'Toads and frogs like that, and snakes. Could we keep a collection?'

'Yes, we could. Yeah.' I wasn't sure. This was my own passion. Did I want to share it with Microscope? He was looking at me. 'Yeah,' I said. 'Yeah. I could show you some places.'

Driving around a bend, I come to a sight that shocks me. For a moment my headlights reveal it. Torn bodies are lying all over the road. I can't stop; I am carried towards them. Tires have skidded on them, pulling them up and crushing them back. Scraps are littered everywhere, pale flesh on the black road; no longer recognisable, just flesh. Among the bodies, live toads are trying to creep through.

These sights confront us with our own indifference. People do try to help. There are volunteer toad patrols, warning signs, tunnels in some places. Occasionally, councils even close minor roads for the migration. A surprising amount of inconvenience is accepted, one might say, for an animal of no commercial value that was once believed to be evil and perhaps many still see as noxious. But the carnage goes on across the country every spring. Any lengthy car journey after dark will show it. We do not, collectively, care enough to make it stop.

What sense of danger do toads feel, as the wheels approach? This thundering object is outside their scale of perception. It is shaking ground; it is a roar. What hits them might as well be an asteroid. Is the pond-music strong in their heads to the very last second? There is consciousness. Then it is gone.

A car crash is like being seized by a hawk. The steering jerks so hard that your hands are thrown off. There's a bang, and a crumpling. These sounds seem far away. Weightless, you are lifted and flung. What were you thinking about, just now?

At thirteen, I was knocked off my bike from behind. I sailed through the air, past my two friends cycling ahead. Flying at about the height of their faces, I turned as I passed; saw the faces begin to show shock. I wanted to wave. Look at me. Now, when I recall that sensation, I see that these could have been the last moments of my consciousness. I might have been saying goodbye to my life. My fancy is that some deep, detached part of me understood the possibility.

I have no memory of hitting the road. Not hurt, I got shakily to my feet. Where was everyone? A man ran up. 'God', he said, 'God.

That's the closest thing I've ever seen. Young man, that wheel missed your head by an inch. I'm an off-duty policeman. I've never seen anything like that.' He was shaking his head. 'He swerved away. I don't know how he missed you.' Someone led me off the road. People gathered round, from the car that had so nearly crushed my head. My friends watched from the edges, kept out by the presence of so many adults. 'Shall I call an ambulance?' asked someone. 'I'm all right,' I said. 'I'm all right.'

Couldn't we skip all the reacting? That was my feeling. We know what happened. No one's hurt. Can't we leave it? I didn't want to talk. We had been cycling to a place we called the Bunnyhole, a tunnel under the railway, where, if you climbed the wall at the opening of the tunnel, you could get through a hole in the wire fence and onto the bank. We sometimes found Slow Worms up there, under an old carpet. I still wanted to go. 'Can't I ride pillion with you?' I asked a friend.

'Better get your bike home now.'

Forty-five years ago, it was. The memory doesn't come to me very often, but whenever it does, it shocks me cold. I can't believe I was so lucky, and I can't believe I came so close. Everything nearly ended in that moment, but after all it was nothing; made no difference, a non-event. Easy to forget it happened. An inch, he said. Was he exaggerating? I ask myself this, as if six inches or a foot would be much better. And I think I *would* feel better. Somehow, the world would seem safer.

My bike had a wheel almost folded in half. I hid it in the shed so that Dad wouldn't see, and secretly bought a new wheel, which Ade helped me put in at his house.

Catriona Sandilands, the feminist nature writer, is moved by the way in which memories are laid down as physical pathways in the brain. 'A memory,' she says, 'traces an electro-chemical pathway from neuron to neuron (called an engram); no two memories follow the same path, and the more often a particular route is followed, the more chemically sensitive particular neurons become to one another. I find this idea quite extraordinarily beautiful: in the act of remembering something, the world is, quite literally, written into our brain structure.'

Sandilands is writing about her mother's Alzheimer's disease. In Alzheimer's patients, these memory-pathways are blocked, it would seem, by deposits of proteins. And when anyone dies, for any reason, the configuration is dissolved – crushed, torn or softened away. The physical paths disintegrate. They are the me-ness of me, the orientation that I have towards all that I know – the people I met by chance and loved, the joys, griefs and angers, the places where things happened, the things that happened but would not have done had I arrived a moment later, the attitudes and fears. The set of pathways constitutes the self, the individual, the unique node in the system. They are what is lost, as the body disperses and joins other forms.

Talons descend. A car bears down. The blow comes from

nowhere. Wildness is a state of exposure to sudden ending. Road accidents are wild.

I drive over the toads and on. Nature tonight is like a First World War field marshal sending men over the top. Three months ahead, on a warm June evening, baby toads, newly metamorphosed, will march out of the pond. Up the bank they come, into the grass. Wherever you look there are hundreds. They are wobbly, on new tiny limbs. These toadlets are still half-tadpole; smooth-skinned, not shaped like toads yet. They are glistening black, dusted with gold. Through the forest of grass they advance. You can't come near the pond without treading on some. A dog runs in and licks them up by the mouthful, oblivious to the toxin. Probably he will be sick. Gulls squabble over them. Jackdaws strut about, gorging. Everywhere there are more: a pixie army.

In 1984 and 1985, a herpetologist, Chris Reading, marked all the new toadlets he could find, as they emerged from a pond in Dorset. The two years combined gave him 7,259. He marked them by clipping a toe from each toad. The operation was carried out 'under licence from the Home Office', and he used 'fine watch-maker's forceps'. It is a relief to know that. Still, these toadlets are so tiny. Moss to them is thick foliage. Their toes are barely large enough to distinguish. How did he restrain these fragile, agitated creatures that a finger and thumb could crush almost without pressure? What he did with the toes is not revealed by the *Journal of Zoology*.

The purpose of the study, along with a similar study in Sweden, was to find out what proportion of the toads that reached breeding age would return to the pond of their birth, rather than go to other ponds within range. Evidently our collective need to know this was considered more important than the need of those seven thousand individuals to retain their full complement of toes. All the ponds nearby were surveyed each year. Of the males that were found, 96 per cent were at the original pond; of the females 93 per cent. But to me the much more striking result is the total number that returned to any pond at all. Surveys between 1985 and 1989, of toads returning to all the ponds, found only 40 of the 7,259 (35 males and 5 females). Of the whole two-year brood, only 0.5 per cent lived to reach maturity, as far as the researchers could see. Less than 1 per cent of that pixie army got through. It is only one sample, but there is no reason to think the rate untypical.

Trevor Beebee, the leading British herpetologist, gives a slightly more optimistic figure: a 5 per cent survival rate from toadlet to breeding adult. In one spawning, a female Common Toad produces between 800 and 2,500 eggs. Just 1 to 5 per cent of these will reach the stage of leaving the water. Of that 1 to 5 per cent, Beebee estimates that 95 per cent will die before sexual maturity. There are records of Common Toads living forty years in captivity. Wild ones aged eight or nine are sometimes found. But they are very unusual. At any given moment, reckons Beebee, only about half the wild adults will survive another year. The larger an animal grows, the more vulnerable it is – conspicuous and cumbersome, though perhaps now too big for some predators. That beautiful large female there, with patches of reddish pigment: will she return next year? She has a 50–50 chance at best. How, under these circumstances, could perception of time evolve? It would be too cruel. When we encounter wild nature by chance, we don't see this impermanence, usually. When a creature is old and scarred, visibly slowing, or sick, then we see it. But most small creatures don't live long enough to reach that condition.

Traditionally, one of the most important distinctions people make between human beings and animals is that we think about the future and animals don't. They have varying capacities for memory and for recognition of individuals and places, but our assumption is that they don't worry about what might happen tomorrow or next year. W. B. Yeats expresses this view in his poem 'Death':

Nor dread nor hope attend
A dying animal;
A man awaits his end
Dreading and hoping all

According to this tradition, an animal has its consciousness extinguished, but it does not face *death*, as human beings do, since the idea of death is not merely that of the physical shut-down and loss of consciousness. Death, as a concept, for human beings, includes the expectation and fear: the knowledge we have throughout our lives that we are going to die. Animals are thought to lack that knowledge. This is why Yeats asserts that 'death' is a human creation. Death often takes us by surprise, but even when it does, it falls upon people who know it will come some time, and have known this since the gradual recognitions of childhood.

That knowledge is central to the sense of self we have. Our self is an entity that is mortal and will, we hope, have a long life, as human lives go. We define ourselves, our hopes and plans, on the basis of that span. When we say that animals live by instinct, we are distinguishing their seasonal behaviour and response to immediate stimulus from the projective fears and plans that feature in our own experience. Neuropsychologist Robert Ornstein expresses the difference like this: 'Being conscious is being aware of being aware. It is one step removed from the raw experience of seeing, smelling, acting, moving, and reaction'.

The distinction sounds like a clear one. But locating the boundary that this one step has to cross, or deciding whether the boundary has been crossed, is famously difficult. In 2012, a group of leading cognitive neuroscientists, neuropharmacologists, neurophysiologists, neuroanatomists and computational neuroscientists produced 'The Cambridge Declaration on Consciousness', which stated that the physical processes associated with consciousness in human beings were also to be found in a wide variety of animal species. The neocortex – the 'new rind' or 'grey matter' that forms the outer layer of the upper part of the brain, possessed only by mammals, and squiggly and immensely complex in primates and human beings

– is not the only 'neurological substrate' that generates consciousness. There are other neurological generators of emotional states and intentional behaviours. Neural circuits that support behavioural and electrophysiological states of attentiveness, sleep and decision-making have been found in insects and in molluscs such as octopuses. There is 'convergent evidence' that 'non-human animals have the neuroanatomical, neurochemical, and neurophysiological substrates of conscious states along with the capacity to exhibit intentional behaviours'. Consciousness is not an attribute confined to human beings or mammals.

But what exactly is meant here by consciousness? Does it necessarily include the time-perspectives that enable a creature to survey its past life and thus conceive of its own identity, becoming 'aware of being aware', and aware that its lifespan will end? Or has the boundary been moved? Can the definition of consciousness include states in which there is only the present, the most immediate reception of sense-messages: no memory, apart from the familiarity of a feeling, no expectation, apart from the way a sensation draws you or a fear propels you? How difficult it is to find words that are clearly one side or the other of that boundary. Here I have used the word 'fear', which implies more than a conditioned response to certain kinds of movement – it implies that the fearful consciousness has an idea, a picture, of the fate to be avoided. Recent studies of dolphins have discovered that captive individuals respond distinctively to recordings of the voices – the 'signature whistles' – of individuals they were once kept with but haven't encountered for twenty years. Dolphins thus seem to have long memories, but do they have conscious memory – the replaying in the head of particular scenes, and thus the construction of themselves as protagonists – or only the unconscious storing of a particular sound pattern? Or something in between?

I don't know whether to imagine these dolphins as seeing their own lives as narratives – bounded sequences with a point of origin and an order of events. Or do they live in the present, not looking to horizons in past or future, remembering that familiar sound – or responding to it, anyway – only when it is played to them in the present? Marcel Proust, the great novelist of memory, valued involuntary memory for its power to take us by surprise, in the sudden recognition of a taste, sound or smell, forgotten by consciousness but stored somewhere in us. Coming all at once, that sort of memory evaded the control of conscious recall, with its already established narrative, full of biases. Perhaps those captive dolphins experienced the dolphin equivalent of a moment of Proustian recall, so powerful because so unexpected. But Proust did not conjure up a state in which involuntary memory was the only kind, a state of having no familiar narratives to ward off, a state of nothing but present sensation, no picture of past or future, no self as protagonist.

Dolphins are sophisticatedly social animals, communicating with each other using subtle signals that identify particular individuals. Toads, as far as we can see, are not. Apart from the way the pairs spawn, toads display no power of co-operation with each other. It is inconceivable that they communicate danger to each other. A toad will surely be disoriented and stressed by the strange sensory world around it on the road. Huge movements and vibrations are all around it; incomprehensible rushing objects in front and behind. There is a shock of exploding air each time a great dark thing goes past; there is spray; there are strange caustic smells. Blinding, impossible lights come and go, followed by darkness and thundering, rattling noises overhead. Stones on the hard road surface jab. All the time, through all this, the pond calls. Skin toxins and blood smells in the air are a warning, perhaps, but the toad does not know what has happened to the others whose bodies it crawls past. It has no picture in its mind of what will happen if the wheel comes down upon it. Not one toad refuses to cross the road, or sits trembling at the edge, wanting to cross but unable to force its limbs.

With other dangers, more natural dangers, such as buzzards, herons, grass snakes, badgers and other predators, not even the sensory stress is there. Like most wild creatures, the toad is fearfully exposed, but it does not live in fear of being seized. Until there is threatening movement immediately ahead – when the toad will stand on straight legs and fill itself with air to look bigger – or until the touch of talon or teeth, the toad is not apprehensive. This is what we mean when we say, in pity and envy, that animals live in the present and live for the moment. We envy, because we identify this living for the moment with living wholeheartedly: not holding back but committing oneself completely, and thus experiencing the full intensity of our senses. An old Romantic theme is that we have traded this ability for the security and leisure civilisation gives us. We cannot repent of the bargain, honestly, but we can feel wistful. People who love watching wild nature are seeking an outlet for that wistfulness, it is often said; as people who love danger are seeking a way of returning for a moment to that animal state.

Perhaps so – this seems very likely. But for me, the relationship between my love of wild nature and my feelings about home is very complicated. I love *watching* wild nature. That means having a conscious sense of being outside it as I watch. I am not there; I can't be touched. 'It is pleasant to be afraid when we are conscious that we are in no kind of danger,' Virginia Woolf said, writing about ghosts. I would add that we are only ever able to be *relatively* conscious that we are in no kind of danger. Really, we know that we are always in danger from the larger world and the future we cannot see. The pleasure is in finding a way to contemplate that truth while feeling detached because we are in no *particular* danger.

When I was young, my love of expeditions into wild nature – my love of the *thought* of wild nature being out there – was an escape from a home where I knew I was loved, yet where I was often frightened and angry. It was an escape from school, too, where I always felt clumsy and nervous – always looked-at, always on the point of embarrassing myself. Wild animals didn't look to me for anything. I had to make no performance for them. Watching them, I saw vulnerability, power and violence – the drama of life, in forms that didn't touch me directly and therefore *could* touch me emotionally. My sense of safety enabled me to feel.

Then, as now, I liked to watch wild nature while thinking about home: to have my attention jerked away from thoughts of home, and then to return to those thoughts. It is about crossing from one to the other, all the time. I like to sit in safety thinking of wildness out there, and I like to be in that wildness while remembering that home still exists, that I am not exposed like these creatures. For I know that my safety is only a greater degree of safety. Talons can descend at any moment. A wild creature's vivid colours – they are all vivid, even the earthy, grassy colours – express the absolute commitment it has to its time and its place. I feel I am always in more than one place, except for flashes of urgency. Wildness is always over there, not here. This makes me feel safer, and in feeling safer I can think about my own exposure.

Chapter Three
Common Frog, Marsh Frog, Edible Frog,
Pool Frog, Smooth Newt, Slow Worm
and Great Crested Newt

———

Frogs turned up in the pond the following spring. As I arrived at the pond one bright February morning, there was a splashing. Parting the weed with my fingers, I saw the frog's striped legs kick out on the bottom. My hand plunged in and came out full of frog. I felt the head butting inside my cold hand, and the legs attempting to straighten. With finger and thumb, I made an opening, through which the frog thrust its head and then its arms – thick forearms: a male. He was muddy green on top and white underneath. His rubbery gape had that slightly unhappy, apprehensive look frogs often possess. After all, he had just been snatched up, most likely by a predator. On my wet skin were specks of black mud and small fragments of weed. The long legs dangled from my hand.

I took him to the middle of the lawn, to see him jump. Straight away he was off, faster than I expected. Before I knew it, he had reached the border and was under the rose bushes. I had to run up the lawn and back down the path on the other side of the bushes, thinking I had lost him, but in under the roses was a movement, and he landed on the paving stone in front of me, enabling me to crouch and grab him from the front as he swivelled to avoid my hand. This time, when I made the hole with thumb and finger, he immediately lunged through it, and my hand closed on his legs. I found I could hold him better like that, with my hand around his bony legs and the rest of his body jerking outside my fingers, arms wide open. He was slippery, though, and was trying to work himself out of my grip, while I tried not to hold him too tight. His back strained with the leap that wouldn't happen, his effort accentuating the sharp angle at which a frog's back turns down. The angle provides the catapulting power.

Back in the pond, he sat blinking for a moment and leaped into the air. Landing on the rim, where grass had covered the plastic, he turned sideways and plopped into the water. I was thrilled.

'There's a frog in the pond,' I announced at breakfast. 'A male.'

'Lovely,' my mother said. 'Just what you wanted.'

'Will there be any spawn?' said Cathy.

'I don't know. It's only the first year.'

'I hope we get our own tadpoles,' she said. 'I'd like that.'

I hoped so too. Since the fish had wrought their carnage and disappeared, the pond had seemed sadly empty. I wanted to see it wriggling with tadpoles. They would be constantly on the move, working their mouths along the leaves and stems of weed and on the plastic sides, which were already thinly coated with algae and no longer looked quite like plastic. Masses of tadpoles would mean movement everywhere in the pond. In every little space there would be something happening.

Next morning, to my excitement, I heard splashing before I got to the pond, and my hands in the freezing water felt several bodies, including a plump, scrabbling pair. The female's stomach was a ball stretched very tight. I felt it, and the four legs pushing at me. The male couldn't steer the heavy female easily. On the soft mud, my hand felt its way under the female and raised her gently. She was as heavy as an egg. Her tight skin had small grainy pinpoints that rubbed on my hand. I brought the pair to the surface, beached on my palm. They tried to find their way off, the female pushing outwards with her arms as if she was swimming, but my hand formed a cup around her. After a moment, the frogs were still. This strange island that had risen beneath them was not like the grab of a predator. There they were – a pair of frogs in *amplexus*, the term used in the books to describe their mating embrace. I had found them in my own pond.

The female, quite a small one, was a rich reddish bronze on her back, flanks and head. Her egg-tight belly was yellow with green marbling. Behind her eye, the dark brown patch stood out boldly. All Common Frogs have this patch. It covers the flat disc of the eardrum, and curves down sharply to the end of the mouth. The dark green male was a large one for this female. His body was as long as hers. On the tight globe of her body, his loose skin quivered. The other males had surfaced and were watching.

That reddish colour, my friends and I later decided, was the best colour a frog could be. Muddy colours were much more common. Some frogs were yellowish with black blotches on their backs – quite pleasing, quite quintessentially froggy, but the blotches looked like some kind of stain, or dead skin that might be spreading. They were too perfectly, uniformly black. Some sort of disease was indicated, some vulnerability, so we thought – and vulnerability often seemed to be the strongest idea to be had from the frogs that we caught. We were wrong about those blotches. They are a common pigmentation, nothing more. But the reddish brown was the colour we looked out for, and it was rare. The books seemed to think it was the classic Common Frog colour, too, for the illustrations mostly showed frogs of this colour.

Now she moved, and found purchase on my hand with a back leg, swivelling herself and her rider. This time I let them tip into the water. In the pond they moved awkwardly about, and then paused, looking out of the water, the male's head almost exactly above hers, the two throats pulsing. Male Common Frogs do not produce a loud croak, but a low bubbling sound, which now started up. I heard one long trill. It sounded as if it might rise to a whine of excitement, but then it stopped. Searching for the source, I saw that one of the males had his throat puffed out a little. But he did not call again. Perhaps I had disturbed him. Mainly they croak at night, when mating is getting frantic.

Rana temporaria, the Common Frog, is the familiar frog seen in gardens; it is the only one found all over the country, and, strictly speaking, the only surviving native frog in Britain. Technically, there is another native frog, but there is only one place where there is any chance of encountering that frog, and it is a secret place. To explain, I will need a digression.

Scientific reclassification has made the roll-call of British frogs more complicated than it was then. In those days, there were three clear species. In addition to the indisputably native Common Frog, there were two introduced frogs, larger and much rarer. One of these, the Marsh Frog, was revealed in the illustrations as a large, broad, muscular plain green frog, with a sharper nose than the Common and bulbous eyes close together on the top of its head. A thick raised vein ran from each eye down the side of the back. When I hunted Marsh Frogs as a boy, their name was *Rana ridibunda*, but now it is *Pelophylax ridibundus*. Either way, the second word means 'laughing', and perhaps the frog laughs at our efforts to name it, among other things.

British Marsh Frogs are descended from twelve specimens placed, apparently out of a passion similar to my own, in a garden pond in a village called Stone-in-Oxney in the Romney Marsh area in 1935. The numerous steep-banked dykes and ditches of the Marsh were a perfect habitat, and the frogs spread rapidly. Twenty years later, they were common across that whole ecosystem. They have since turned up in other marshland areas with drainage ditches, in northern Kent, Sussex, Somerset and West London. No corridors of easily-crossed habitat connect these places with Romney Marsh or with each other, so it seems certain that people have been spreading these frogs deliberately – again, presumably, out of a romantic desire like the one I shared with my friends for a landscape full of strange exciting creatures. Or perhaps people took some home as pets, and their frogs escaped. Living in a clearly defined region within cycling distance of London, in the ditches and channels visible as soon as we entered that region, Marsh Frogs proved easy for us to find.

The third species was much harder to find in the flesh, and we never did, despite following several leads. In the books we saw a plump frog of similar shape to the Marsh, but more highly coloured – a brilliant light green, as if painted and glazed, with black blotches and lines. A yellow line ran down the back. The belly, the sag of skin under the mouth and the insides of the legs gleamed creamy white. This was the Edible Frog, then called *Rana esculenta*. We read that there had been many releases of this frog in southern England across the nineteenth and early twentieth centuries. Colonies had survived in several places, but were scattered and localised. Nowhere had they merged into an extensive population. On anecdotal evidence, it was thought also that the Edible Frog had been in Britain before the nineteenth century. The Romans had introduced it, intentionally or not. They were importing one of their staple foods. For the same reason, it was likely that there had been later introductions by Saxons and visiting French soldiers. Across Europe, frogs were an important seasonal part of the basic diet, and had been so since prehistoric times. Any kind of frog could be eaten, but the plump thighs of these large species were especially worth the catching and cooking. The most determined attempt on record to introduce Edible Frogs to Britain was that of a Norfolk squire, George Berney of Morton Hall, who deposited into his local ponds 200 individuals bought in Paris 'and a great quantity of spawn' in 1837, followed by another consignment from Brussels in 1841, and in 1842, 1,300 frogs from St Omer. Berney's motives do not seem to have been recorded.

Yet the results of this long history seemed to be patchy and precarious. Edible Frogs could breed in this country, but most of the colonies had faded away in a few generations. British Edibles might be on the way out – that was the message. Here and there, the species was surviving in a pond or two, but nowhere had they spread widely enough to be secure. Something made it difficult for Edibles to 'take', whereas in only 30 years, and from only 12 individuals, the Marsh Frogs had become a boisterous multitude.

In the books we loved as children, there was no mention of another continental frog, the Pool Frog, then called *Rana lessonae* and since 2000 *Pelophylax lessonae*. This frog is similar in appearance to the Marsh and the Edible, with the same sharp nose and eyes like marbles on top of the head. It is darker in colour, brown or green, and has the dorsal yellow line. Pool Frogs are found from France throughout central and southern Europe, into Russia and up to Sweden. Until the 1970s they were classified as a subspecies of the Edible Frog. Genetic science then revealed that the Edible was not a species in its own right but a hybrid produced by the interbreeding of Pool

Frogs and Marsh Frogs – hence the addition of *klepton*, meaning thief, to the scientific name. A klepton is an animal that in order to reproduce needs to 'steal' or 'borrow' additional genetic material by mating with another species. The Edible Frog is now *Pelophylax klepton ridibundus*. It is a fertile hybrid, but fertile in a highly unusual way.

The primary hybridisation occurred, and still does, through the mating of male Pool Frogs with female Marsh Frogs. It has to be that way round, and for some reason the reverse pairing hardly ever takes place. What has enabled the Edible Frog, unusually for a hybrid, to spread beyond territories in which the two parent species live together, is that, by some genetic chance, not yet explained, the Pool Frog chromosomes inherited by the Edible are not present in the Edible's eggs or sperm, which contain only Marsh Frog chromosomes. Edible Frogs can therefore reproduce by mating with Pool Frogs; this will always produce a mixture of Marsh and Pool chromosomes, and thus a hybrid, an Edible Frog. This is why Edibles were able to spread outside the northern central regions of Europe where Marsh and Pool Frogs coexist. They could go wherever Pool Frogs were present. Thus they spread to northern and central France and northern Italy, where Marsh Frogs are not found. In Britain, the introduced populations that were able to breed must also have included some Pool Frogs – always likely, since any wild Edible population will be living alongside Pools.

Two other scientific developments have occurred since that time.

One is a shift in classification only: the change, for Marsh, Edible and Pool Frogs, from *Rana* to *Pelophylax*. These three species are European green water frogs, different in shape and colour from the Common Frog, and more aquatic. After breeding, they do not leave the water and disperse, but stay in it, venturing at most a few feet away. The frogs bask on the banks and on floating weed at the water's edge, jumping in with a splash when disturbed. Unlike the Common Frog, these green frogs love exposure to the sun. Their habit of remaining for long periods in the water with just their eyes above the surface is responsible for the raised position of those eyes. In 1843, the Austrian naturalist and zoo director Leopold Fitzinger proposed that these green water frogs should not be classed with the brown frogs, such as the Common Frog, in the huge genus *Rana*, where Linnaeus had put them 90 years previously. Instead, Fitzinger wanted them recognised as a separate genus, called *Pelophylax*. For 150 years, most scientists in the field rejected this view, but at the end of the twentieth century, studies of DNA sequences confirmed it.

The other development was research, published in 1999, 2001 and 2002, identifying partially fossilised pelvic bones from Lincolnshire and Cambridgeshire as Pool Frog bones. These bones dated from the middle Saxon period, between AD 600 and AD 950, and their identification was evidence that Pool Frogs, either native or intro-duced, were present in eastern England at that time. In the absence of any tangible evidence of introductions in Roman or Saxon times, the bones indicated the possibility that a native population had survived the Ice Ages. Green water frogs were native to Britain in the Pleistocene era, as fossil evidence from three sites reveals. The general assumption was that glaciation had wiped them out.

Now, with the bone identifications, opinion began to tilt the other way. Anecdotes from before the recorded Edible Frog introductions of the early nineteenth century tell of frogs that don't seem to be Common Frogs – loudly croaking frogs known locally as 'Dutch nightingales' or 'Whaddon organs', though the sounds described may have come from Natterjack Toads. A Pool Frog population near Thetford in Norfolk became the object of intense scientific scrutiny in the 1990s, since the frogs at that site resembled the browner Pool Frogs found in Sweden rather than the greener ones of France and Italy. The Norfolk colony lived in a 'pingo', an ancient pond in a crater formed by the melting, at the end of the last Ice Age, of an underground swelling of ice. These brown Pool Frogs – the northern 'race' or genetic 'clade' – are the ones adapted to life in a climate similar to Britain's, and the ones most likely to have spread into Britain before the separation of the British land-mass from the continent. Most of the other Edible and Pool colonies in Britain, concentrated mainly in the south-east, are clearly of more southern origin.

In the 1990s, the view gained ground that the Thetford frogs, at least, were probably natives, descended from a more widespread pre-glacial population. With the publication of the bone research, this view became the official opinion of the government, as represented by English Nature, the main government agency responsible for the conservation of Britain's natural heritage. *Pelophylax lessonae* is now classified as a native species, within the restrictive definition of that term used by government agencies and written into law. To qualify as native, a species must have arrived and established itself in a country independently of any human activity, no matter how long ago the arrival occurred. Human beings are not, for these purposes, considered part of the ecological process. Since 1981, it has been illegal to release non-native animals in Britain, including individuals captured here, belonging to species well-established here. If you capture a Marsh Frog for any reason, for example, you are not allowed to let it go. English Nature, incidentally, has now been renamed Natural England. The names of government organisations are even more unstable than the scientific names of frogs.

Uncertainty persists. Some herpetologists remain doubtful about the bone identifications, and feel that the evidence for the reclassification of the Pool Frog as a native is far from conclusive. Introduction during the Roman occupation or later remains a strong possibility. The bones that have played such an important role were found close to sites of human settlement. But whatever the continuing debate, this government recognition has had important practical consequences, for it changed the legal status of the British Pool Frogs. No longer were they aliens, not to be released into the wild. Now officially a native species, they were to be protected. Laws about land-use could be enforced in the defence of Pool Frog colonies. Government and EU funds were available for the purpose of assisting Pool Frog survival.

So it is a sardonic irony that the period in the 1990s when the scientific evidence for reclassification was finally being discovered was also the period in which the Thetford colony, the last known population of the brown Pool Frogs that might be native, was rapidly dying out. The last survivors were three males, the only frogs found at the site in 1993; they were calling fruitlessly for a female. One of these males was captured and taken to London to breed with imported females, in the hope that some of the native British Pool Frog genes – if such they were eventually pronounced to be – might after all be passed to a new generation. After that year, no more wild ones were seen. The 'rescued' male died in 1999, about a year before the analysis of the first of the bones was published. So deliciously exact is this timing, so cruelly neat, that I am struggling not to find it emblematic of our wider response to environmental problems: we recognise them just as it becomes too late for effective action. I hope that we, and fate, are not really that perverse.

Local frog populations sometimes do die out, when their ecosystem changes. There are many variables – the state of the water, the silting up of the pond, the encroachment of new vegetation, the arrival of new predators, or a run of adverse weather. To be relatively secure, a population must be spread across numerous connected sites, so that replenishment is a possibility. One of the naturalists involved in the study of the final years of the Thetford colony may have had a sense of irony like that of Samuel Beckett, since the last male, the one they took to London, was christened Lucky.

A new native frog was given to the British people and snatched back at the same moment, but the rules made by human beings about things like this are full of paradox, and although a species that originally came to a place through some sort of human agency is not regarded as native however long it has been present, once it is determined that a species probably arrived without human agency, then human agency can be used to replace it when it disappears. I hope I got that right. Such is the squiggliness of the lines we draw between human works and natural occurrences.

Local extinctions can be reversed, if we choose. In 2005, English Nature authorised and funded the release of 170 imported Pool Frogs at a site close to the home of the extinct Thetford colony. Ponds, including old pingo formations, had been deepened out to receive them. Predator numbers are being carefully monitored. The location is a strictly kept secret. Further releases are planned, though there is anxiety that if this one should fail to 'take', for any reason, funding for subsequent attempts will be hard to obtain. Sadly, the released frogs did not include any of Lucky's descendants, since he was introduced in his London pond to Dutch females, and by the time of his release the Swedish population had been identified as the one closest genetically to the Thetford frogs. So there is no direct continuity. No one knows whether any native Pool Frogs survive elsewhere in the depths of some fen, and indeed whether any since the Ice Age have truly been native, as the law defines it.

I am glad that the reintroduction is taking place, and I confess that I don't much care whether the species was ever native. As yet I am not allowed to go and see them, but I am pleased they are here. Britain has too few reptiles and amphibians for my liking. From the depths of my old childhood enthusiasm, I want more, and I wish that human romantic enthusiasm and ecological variety could be the unabashed criteria, without need for these wobbly distinctions between nature and human contrivance. In the new era some geologists are calling the Anthropocene – the time of the end of nature, in which all ecosystems are being changed by human activity, and the state of ecosystems everywhere must be accepted as a human responsibility – such distinctions are ever harder to justify.

But I understand the position conservationists are in, and the difficulty of the judgements involved. Newly arriving species can cause a lot of unforeseen damage; there is no doubt about that. They can bring diseases to which the present species are not adapted, with catastrophic consequences – look at Dutch Elm Disease and the Oak Processionary Moth. New competitors can drive established species to extinction – look at what the American Grey Squirrel has done to the Red Squirrel in most of Britain. Our desire to preserve the 'native' or traditional species, the ones deep in our culture, the ones we love, is an anti-ecological desire, strictly speaking: a desire to freeze the process and prevent the endless succession that is the life of the evolving ecosystem. If the test applied by conservationists is that we can only increase biodiversity by reintroducing extinct native species rather than adding new ones, then at least they can say that any unpopular ecological consequences are natural.

Frogs do not make good captive animals. Even Microscope and I understood this quickly. Their reflex response to anything that startles them is to leap forward, preferably into water. They need water of sufficient breadth and depth to leap into and hide. In the miniature landscapes we built inside galvanised iron tubs and long baths, this was difficult to provide. We only had flower-pot saucers sunk into the soil and filled with water. Over each enclosure we draped net curtain, tying elastic tight around it under the rim, and pulling the curtain tight, held in place by the elastic. Many of the frogs we caught never seemed to settle, among the grass tussocks we planted and the flowerpot halves we provided for them to sit under. Unlike the toads, they kept jumping against the sides of the enclosures, until there were pink patches on their noses where the skin was rubbed away. We felt sorry and released them.

It was enough to have them in the pond, where, a few mornings after I saw the first male, I found three clumps of spawn, the globes of jelly still small, dense and rubbery. I cupped my hand under one, and felt its weight. The adults were nowhere to be seen. At some time in the night, my small colony might have had a moment of 'explosive breeding'. A female releases her eggs, pressing her hands against her swollen belly to squeeze the spawn out. Like the male toad, the male frog comes to sexual climax at the touch of the spawn on his feet, but in frogs the spawn comes in a rush, spurting out as a thick liquid. The sperm must mix with this liquid in seconds, before contact with water sets the spawn into rubbery jelly. Unattached males scent the spawn and the sperm, and press in around the pair, their croaking bubbling up into a fizz. They move in and touch the jelly. Perhaps other frogs have spawned already, and the males are now slapping and sliding across masses of jelly. Touch, scent and sound bring all these males to ejaculation, all together. The water becomes a soup of spawn and sperm, mixed up by the thrashing bodies. Other pairs are brought to their moment of spawning by the excitement. The male on the female's back fertilises most of her eggs, but some receive sperm from other males, clambering, kicking and pushing.

For a few hours, the jelly is dense and tight around the black dot, the fertilised egg, the *vitellus*. Then the globes absorb water and loosen and swell. Each globe is stuck to several others. They are stuck where they touch but do not form a continuous mass. Water can pass through the clump, providing oxygen. In ponds where there is soft mud on the bottom, sent up in clouds by everything that moves, the globes are soon coated in light-brown dust. They look like submerged bunches of small grapes. Inside them, the developing black embryos do not feed on the jelly and eventually eat their way out, which is what we thought at the time. The jelly is for support, warmth and protection. If it were highly nutritious, it would not last long in a pond full of mouths. In a week to ten days, depending on warmth, the black tadpoles have grown and lengthened, and when they reach this stage, they produce an enzyme that begins to dissolve the jelly. For a short time, in the softening jelly, they have feathery gills to filter the mucous out as they breathe. Wriggling gently, the tadpoles collect in black swarms on the empty jelly. When it breaks up and they swim free, the gills disappear.

At first, the tadpoles feed on vegetable matter, grazing algae from surfaces and sucking in strings of weed. By this time they have plump gold-speckled bodies and horselike faces. Later they need animal food, mainly carrion. Moving the weed aside with my hand, one morning that spring, I saw on the mud at the bottom a neat line of rippling tails and busy bodies. They were nibbling at something white and furry. It was the underside of a shrew that must have fallen into the water. I saw the whole body now, the dark velvet fur. Along the belly the tadpoles were lined up like suckling young. Beneath the shrew's long pointed nose, the mouth was parted in a grimace, showing needlepoint teeth. It was the face of a villainous wizard. I remembered the cards from a Happy Families set we used to play with in my family: Mister Shrew in a top hat, Mrs Shrew in a bonnet, Master Shrew and Miss Shrew in neat school uniform, wearing round glasses. All had sharp noses, small twisted mouths and spiteful teeth. The parents were frightening; the young shrews, earnest in their uniforms, were trying to be normal.

A few days later, when I looked again, there was a perfect clean skeleton in the mud, white and delicate.

Quickly we came to see Common Frogs as too common, not worth much attention. The previous summer, the sight of one, leaping from our feet in long grass, would have prompted us to dive and grab and hold it up triumphant. Now we just pointed and let them disappear. Only a frog of striking colours and markings would attract our attention, and then we would take it home for release in my pond; not try to keep it in our collection. I began to think of them then as I think of them now. It is best to see them or imagine them deep in the wet grass or the brown leaves of a woodland floor, where their sudden leap is a part of the undergrowth coming to life, and their eyes look bright and interested, not apprehensive.

Our collection was growing. Suppressing my reluctance, I had taken Microscope to some of my favourite places. In a large park, a short cycle ride away, we had watched the keeper until he was out of sight and then waded into the pond in the rock garden, heading for the bridge that crossed it. Under that bridge, in the shadow, the pond got suddenly much deeper; there was an underwater shelf. This was where the toads gathered to breed. There was no sign of the keeper coming back. Ducking under the bridge, we bent over, looking into the brown depths. Cold water lapped into our wellies.

Toads could be seen moving on the bottom. There was a light-coloured male, the thick folds of his skin clearly visible.

Micro had brought a large net, which his mother had made with him. It was a huge pair of white underpants with the legs sewn up and a square wooden frame stitched into the waistband. The net was nailed onto a thick pole like a broom handle. Altogether, it was formidable, much more sturdy and capacious than the seaside net I had brought. I was impressed.

'Whose pants were they?' I asked, but he said he didn't know. He put the net into the water and jabbed it into hidden space under the weed. The little male toad had vanished. When Micro pulled the net out, the pants ballooned full of water, in a shape that suggested fat buttocks and thighs. We both giggled as the water streamed away. 'My mum got them in a jumble sale,' he said.

'Look, over there.' I had seen a movement in the blanket weed. There was a flurry of toad legs and bodies.

'Can you get the net right under?' He leaned out, precariously. 'I can't reach,' he said.

'What if I hold your arm?'

'Yeah. OK.' I gripped him just below the armpit, with both hands. He leaned out over the water, almost pulling me in, but I righted myself and hooked my other arm on to the bridge. Micro brought the net round in a great sweep, and then up under the weed. The whole blanket was sucked into the net, which bulged alarmingly, but the pants were strong. Peering in, we saw movement. 'Let's get this net back to the bank,' I said. Micro turned and sloshed his way proudly to the side of the pond. I followed.

We pulled out the yellow-green blanket weed, a great felted tangle, with leaves in it, and snails, and wriggling nasty-looking insects with pincers, and Smooth Newts we would normally have paused over, but we had some of those already. Pulling apart the matted weed, we found what we wanted: four round golden eyes peering out of the green tangle, and a fat struggling body trussed-up by the weed. No, two bodies – it was a pair of toads, the female large and liver-coloured, and the small plain male clinging tightly, his head fitting neatly between her eye-bumps.

Strings of spawn hung from her anus, and I realised, with a moment of guilt, that we had caught them in the middle of spawning. Now I could see, in the weed here and there, lines of black dots wrapped in jelly. Out of the water, the jelly was no more than a smear of mucus on the weed. It looked fragile. In pulling the weed apart, we had broken the strings in places. I put my finger under a string hanging between two clumps of weed, and lifted my finger, testing the strength of the jelly. It parted immediately. Black dots hung from my finger in the sagging mucus: little planets.

'Here comes the keeper,' said Micro, quietly. I looked up.

He was half way across the lawn, walking towards us: a man in green overalls, red-faced with oily grey hair. We had encountered him before, waiting for us behind a bush when we came out of the water. 'Don't go in the pond again,' he had said, tipping our newts back into the water: the lovely purplish male Smooth Newt that had excited me so much. 'It'll be the police next time.'

'Oy,' shouted the man, when he saw we had seen him. 'Stop right there.'

He started running.

I knew my father would cane me if I was caught.

'Up over the fence,' said Micro. The fence was splintery reddish wood that looked new. Behind it was someone's garden. 'Quick,' said Micro. He was up the straggly tree next to the fence and straddling the lintel. 'Pass me the net,' he said. I held it up by the handle; he took the pants carefully, peering inside. 'Now come on,' he said, giving me his hand. I wasn't good at this, but I knew the man was coming, and somehow Micro pulled me up the tree, the water slopping inside my wellies.

'You little buggers!' the keeper shouted. 'You're in big trouble if you go in there.'

Micro put the net in my hands and dropped down into the garden. I passed it to him, and dropped down myself, almost crashing into some glass in wooden frames. No one seemed to have seen us from the house. We looked around.

'Over that fence,' said Micro. It was an easy one, and it took us into the neighbouring garden. The next barrier was a hedge with wire netting behind it: easy again. Looking cautiously around, we crouched in the lee of the hedge and shuffled towards the side of the house, where a wooden door into the front garden looked climb-able. It wasn't locked. Seconds later we were out in the road.

Amazingly, the toads were still in the net, attached by the spawn to a mass of the weed. We later found we had two Smooth Newts and a water scorpion as well. At first the toads went into a fish tank with our newts, to finish spawning. It was cramped but it was all we had. Next morning we put all the spawn in the pond, but probably it was too squashed and battered. No tadpoles appeared in the weeks that followed.

This pair of toads became favourites for us. The female we called Fats and the male Titch. These were not very original names, but the word 'Fats', in particular, said with a smack of the lips, conjured up hugeness of appetite and bagginess of body. We delighted in dropping woodlice under their noses and watching a toad crane forward out of a flowerpot half. A soft pop, and the woodlouse was no longer there, while the toad was stretching its head strangely, submerging its eyes. Then it sat back, and settled its feet, and was ready for more.

We also had Smooth Newts in an aquarium: three carefully selected pairs, as if we were breeders. One male was a dark purplish brown, one pale tawny and one light grey. All were spotted in black like leopards and had crests with rounded tips, rather than the toothed crests of the Great Crested Newts we longed to find. The Smooth Newt is the commonest newt, found almost everywhere, like the Common Frog, and already we were rather unconcerned with them, and would drop them back one by one as we searched for the legendary Crested. But they are beautiful. Because the male's crest is not jagged, it makes a long, unbroken, arching, rippling curve down back and tail. I always searched for one with a high crest, like the ones I saw in pictures, but often, in the tank, the crests were smaller than they had seemed in pond and jam jar. Still, the males would come stalking through the weed, their striped heads poised and watchful. They would swim a little way, float, and drop slowly to the bottom, where they would rest, one foot raised to part the weed-fronds like a curtain.

A male sees a light brown female. He approaches from behind to touch her cloaca with his nose. If the scent excites him, telling him she is ready, he will launch himself up and swim over her, turning to face her and drumming his tail on the side of his body, so fast as to make the tail blur. I love to see this even now: looking down on the newts in a park pond. At university once, I remember, I was revising for my exams. My room became unbearable. I ran out and started pacing through the gardens, in a state of sick tension, as I tried to learn sentences off by heart. Stopping beside a pond on the campus, I looked down, and saw Smooth Newts drumming their tails. And I smiled, and my body relaxed. There would still be newts in ponds, whatever happened.

The male drums his tail at the plump brown female, flashing the line of iridescent blue or mauve that runs a little way along it from his cloaca. He is hoping to impress and excite her enough to make her circle and approach him from behind. If she does, and comes up to touch his tail with her nose, he comes to climax, lifting and folding his tail. Enlarged and loosened, his cloaca opens to drop a spermatophore, a blob of sperm in a soluble glutinous packet. Now he steps forward carefully, leading her on until her cloaca is just ahead of the packet. Turning sideways he blocks her way, tail folded against his side. She nudges the tail with her nose, making him unfold it while keeping it rigid, so that it pushes her slightly back, and her cloaca, now aroused and open, descends on the packet which sticks to her and is drawn in. It is an exquisite dance, in which the two animals manipulate and respond to each other with excitement and precision.

We had other animals too. At the Bunnyhole, in the midst of a busy suburb, we had found Slow Worms. The first time we saw one, I spotted what looked like a gleaming iron ring, half-buried in the pale brown grass: just an inch emerging, curving and disappearing. I touched it. The surface was smooth and warm, and though firm it was not hard like metal: it was alive. And it began to move, sliding into the grass roots. I held it between two fingers and jerked my hand back, pulling out of the grass a slim shiny pale grey creature that started whipping frantically in my fingers. It had no neck; its head was no thicker than its body. I ran my fingers closer to its head, which stopped the whipping. What did I have in my hand? I took a close look. The nose was bluntly rounded. Small eyes were black and gentle. A dark tongue appeared, very slightly forked, and licked around the mouth, where there was shiny black stippling.

Beneath a mattress in the brambles we found another – light bronze this time, with a black line down her back. Her sides and face were black and glossy, with faint gold patterning. When we looked them up, we found that the first was a male and this was a female. Her tail ended abruptly, with no taper but a short dark cone, coming to a point. She had dropped it, and this cone was all that had grown from the stump. *Anguis fragilis* is the scientific name for this creature: the fragile serpent. They drop their tails readily, and tail is hard to distinguish from body on legless lizards.

We kept the Slow Worms in a glass tank, between the Smooth Newts and the baby Red-Eared Terrapins we were always buying from the pet shop, though they never lasted long.

Once, at the Bunnyhole, we saw a lizard, wood-brown with dark and light markings. I was daydreaming, not really looking, and there it was, on the edge of the mattress, its long tail curving round. I froze. 'Micro,' I hissed, hardly daring to move. 'Come here.' He was further along the bank. 'Come here, quick.' He looked where I was pointing and gasped. 'Bloo—dy hell.'

'Don't move,' I said.

I reached out with my hand.

The lizard ran the length of the mattress and was gone. I caught a flash of orange belly. We lunged and turned over the mattress. Slugs and vole trails were revealed. The lizard was nowhere to be seen.

'Was it really there?'

'Yeah, Micro, it was. We've got to get it.'

But we never saw one there again, though we came looking many times, turning the mattress over and over, and hunting for hours on the stony, brambly bank as trains roared past above us. When we told other boys, they nodded wisely. But they probably didn't believe us. It had been such a split-second flicker, we weren't sure ourselves.

What we really dreamed of having was a pair of Great Crested Newts. This was the rarest British newt, much bigger than the Smooth and Palmate, and the thought of it besotted me. They were black as fresh tar and had deep jagged crests. Their bellies were brilliant orange with big blotches. Old books from the library called them The Great Crested or Warty Newt, or just The Warty Newt, for their skins were a surface of pimples, like black caviar. Their golden eyes gleamed.

I never saw one, not then, when I was hunting the ponds and wastelands near home with Micro, nor later, cycling into the country with Adrian and Phil, who reported once that a boy at their school called Tony Luffingham had been boasting that he could get them. Weeks passed and he never brought any. Someone talked about a pond down in Sussex, and we cycled out of London with vague directions. We found a pond, perhaps the right one, and waded about until it was muddy soup and we had dragged every clump of weed onto the bank. Silently, we picked through the weed. A dark female

Smooth, fat and squashy, was caught in the tangle; it was a noose around her body, too tight. I eased the loop over her head. She was the biggest Smooth I had ever seen, plump with eggs. In my fingers she flapped like a fish. We said she was a Crested, but she wasn't.

Sometimes, when we searched in a pond, I thought I glimpsed one, out in the middle, in the ravines of weed. But I could never be sure.

Recently, I remembered that expedition, as I stood gazing into a pond I know well. How we fantasised about this creature; it was so powerful in our imaginations. Now I know how to find this newt if I want to. From February to the end of May, they will be there.

Triturus cristatus, the Great Crested Newt, has a different place now in our culture. It is highly protected – perhaps the most famously protected of our rare species. Developers dread it, since the presence of these newts mandates the provision of an alternative habitat and the relocation of the newts before building can take place. In York, in 2014, building work on a new John Lewis store was held up for three months, because Great Cresteds were found on the site in unexpected numbers. All the newts had by law to be captured by handlers with licenses from Natural England, and released in a new wetland nearby, constructed at a cost of three hundred thousand pounds. A small industry has grown up, offering newt surveys and special newt-proof fences. On construction sites, narrow pathways fenced on either side by half-buried plastic sheeting have become familiar. These are the identified migration pathways for the newts, and in spring when the newts are moving, licensed handlers with torches patrol these routes trying to catch every one.

Local headlines scream about the cost of all this, the amenities delayed, the jobs endangered; all for a few small newts, though it is notable that the same writers, when they come to describe these animals, become lyrical. The newts are majestic. They are tiny dragons. When they actually think about this creature, even the people who want to call it unimportant seem to feel a kind of awe. Or perhaps it is just that this kind of writing is what the journalists find in the archives and have to use, though it jars with their indignation.

All this effort suggests that we love these newts – that they roam through our collective imagination, like small dragons indeed – as they did in the 1960s for me, Micro, Adrian and Phil. But the odd thing, the paradox, is that in protecting these newts so fiercely, we have barred them from having the presence in our lives that would bring them into our imaginations. Children catching them now would be breaking the law quite seriously. Some conservationists argue, for example, that shining a torch on these creatures in their breeding ponds at night – the only reliable way to observe them and certainly the only way to see their spectacular breeding displays – might count as disturbance and thus be illegal. It is a hotly contested view.

Three years ago I saw the dance of the Great Crested Newts. I am glad that I did not know then that watching by torchlight might not be allowed.

In the morning the pond had looked drab and empty. Reeds were still brown from the winter. Crinkled leaves floated. I had been told there were Great Crested Newts, but was not very hopeful. It was too early in the year, and too cold. I walked around the pond. Beetles darted here and there. A leech lengthened and contracted like a tiny elephant's trunk. Something about two sticks on the bottom, one across the other, made me look again. Thick little muddy black sticks – but I saw it now, the line of crest running down each back, the blunt heads, transparent jelly round the eyes. Even the feet in the mud were clear now, with the toes ringed in yellow. Two male Great Crested Newts, motionless.

I crouched and leaned forward. A puff of mud and they were gone.

After dark I went back. Spots of rain pricked my cheeks. My beam on the weed found a toad, a loose bag of a body, now emptied of spawn. It tipped forward and swam jerkily, butting the bank. Small newts hung in the water.

There was a white shape over there, a carrier bag billowing, and next to it a male Great Crested in full display, arching his back. He twisted and corkscrewed, flashing his belly, yolk yellow blotched with black. Where was he looking? At the edge of my circle of light was the female, flat-backed, spongy. She walked towards him. He made a little leap, to land in front of her, jagged crest quivering.

Large and deliberate, she pushed past. He swam alongside. She pushed on. He launched himself to get ahead of her, turned, put his nose to hers, and let tail and body lift until he was dancing vertically, nose down, a writhing coil of black and yellow. She was still. His eyes were mad like a bull's.

A smaller male appeared, swimming to catch up, and as he turned, I saw that a leg was missing, or the flesh at least. Just the bone stuck out, white and stiff, like an ivory toothpick. I could see the joint. Something had pulled off all the tissue like a glove.

The female moved on.

For most of the year they live out of the water and tighten into drab, slow, rubbery lumps. Dust sticks to their bodies. The crests have gone. But in spring, in the ponds, those bodies soften and open out, like paper flowers in water, or sea anemones when the tide returns. The males begin to dance, their crests flaming. What does it feel like, that loosening and frenzy? What do the females see, when this mad, bright creature coils twisting in front of them? They dance at night. The whole pond is darkness. Something must seep in through her porous skin, her froth of skin, softening her, filling her.

I switched off my torch. I had rain in my hair. A car went past. Houses were all around. Did people here know about these newts? Did they throng the bank when the season started, crisscrossing the pond with their torchbeams, whispering, pointing. Why didn't they? It could be like the cranes in Nebraska.

Back at my car, under a streetlamp, something on the ground

caught my eye. A female was poised on the edge of the kerb, gazing out like a gargoyle, flexing her toes, sensing the pond, tense at the huge things around her. My foot was inches away. What could her eyes and skin do with these lights and cruel tarmac? Once I would have gasped at this find. I carried her to the pond, cold in my hand, and to drop her in turned on my torch. The male was still dancing.

'Two boys at my school have got a Grass Snake,' said Micro.

I didn't like him talking about his school. We were twelve now, and I had been sent to a different school, a fee-paying school an hour away on the bus. None of my friends were there. The main building had wide steps and stone columns. On the first morning, all the new boys waited on the grass below those steps. Mothers and fathers waited with their sons. My father had come in with me. I had a new brown brief case, but the others seemed to have black ones. A teacher on the steps said it was time for the parents to leave. 'You'll be fine, Rich,' said my dad. 'I'll see you here at four.' I watched his bald head go down the drive. We waited, in the smell of the trampled grass, until our names were called out and we were led in lines to our classrooms.

The teachers wore gowns and were called Masters. I was in Moffat House. Homework was called Prep. At break we could go to the Tuck Shop. I stared at the depth of scratched words on my desk.

All around the school, there were playing fields. Beyond them, the houses looked a long way away, like a distant town. But in one corner there was long grass, nettles, elderflower bushes, an old shed. There could be frogs in there, I thought; maybe Slow Worms.

At the end of that first week, I was standing in the playground – they called it the 'quad' – feeling lonely. I was outside the library, looking in at a dark high-ceilinged room with tall book-cases. Boys were sitting at desks. At the far end was a master, at a table.

The boy nearest the window was doing something under his desk. He was an older boy, maybe sixteen. There was a ball of grey fluff. I couldn't quite see what it was. He had a wastepaper bin at his feet. Under the desk, the boy stretched out his arm, and I saw that he held a large penknife, like an army knife. He now turned sideways in his seat, and I saw that he was preparing to skin a grey squirrel.

Fascinated, I watched him slot the point of the knife in, under the jaw, and draw the blade down the white belly, spilling the gizzards, which he shook into the bin. No one seemed to see what he was doing, or no one was concerned. Perhaps it was expected at this school. He cut off the head, and dropped it into the bin. Then he cut a bit more, and began to peel back the skin. Its inside surface was white and veined. Holding the animal over the bin between his knees, he pulled off the whole pelt, letting the pink body drop like a banana from its skin. The boy was wiping his hands now, and looking furtively around. His eyes met mine. I turned away, fast, and hoped I wouldn't meet him in the quad.

'Their names are Copper and Loader,' said Micro. 'They're in the next year up. They've got a Grass Snake and loads of frogs and toads. Steve Rose's brother told me. They've got an eel.'

I rang Micro's doorbell every evening that week. He didn't seem to mind, but didn't look pleased, either. 'What's your new school like?' said his Mum. 'Is it really posh?' 'I hate it,' I said. The second night, as I went out, my dad stopped me. 'Haven't you got home-work?' 'I've done it already,' I said.

We hung around in the allotments and the park. These places are still here for me, I was telling myself. I can still see Micro in the evenings. I'm not going to be swallowed by that school.

'It's them!' Micro said. 'Over there.'

Two boys were walking about in the long grass by the fence. One dropped to his knees and threw himself down. He was fumbling with something in the grass. There was a jam jar in his hand. We walked over.

'Hi,' said Micro.

The boy with the jam jar looked up. I could see a grasshopper struggling in his hands, its legs waving. His hands were freckled. 'Hold on,' said the boy. He crouched, gripping the jar between his knees, and unscrewed the top with his free hand. I could see grasshoppers inside, leaping against the glass walls. He lifted the lid just a slit at one side, and carefully inserted the grasshopper he was holding. Screwing the lid back on, he stood up, a muscular boy with a freckled face and light brown wavy hair.

'I'm Micro. I saw you at school. You're Philip Loader, aren't you?'

'Yeah.'

'And that's Adrian Copper.'

'Yeah.' A taller boy, also freckled, with very curly, bouncy, dark chestnut hair was walking towards us. 'Hiya,' he said.

'This is Richard Kerridge,' said Micro. 'We collect animals too.'

'Oh yeah?' said Adrian.

'What are you doing?'

'Catching grasshoppers for our toads.'

'Can we help?'

'Hmm. I suppose so.'

'And then can we see your collection?'

We went back with them to Adrian Copper's house, where his bedroom was full of glass tanks, on tables and bookshelves all around his bed. There were Smooth Newts, Palmate Newts, Slow Worms, Sticklebacks, two kinds of terrapin and a large tank that drew my eyes from the beginning. A light-bulb in the aluminium lid cast hot yellow light onto a plant with thick leaves. Droplets clung to the insides of the glass.

I ran over to it. 'Wow. You've got tree frogs.'

They were European Tree Frogs. I had seen pictures of them,

small frogs with smooth sticky bodies, a gleaming light green, a single pure colour. Six of them, close together, adhered to the glass in one of the top corners, eyes closed. In the plant I found two more, one of them climbing from one branch to another, wrapping arms around the upper branch and pulling itself up, showing its pinkish white belly, a froth of pimples. These frogs had looked beautiful in the pictures, but what I saw now was the texture of those glistening bodies, the way skin unpeeled from skin, thigh from belly, clinging for a moment then sliding away. Their toes ended in tiny adhesive discs. I had marvelled at the pictures of these frogs, thinking how joyful it must be to live in a country with creatures like this. There was one colony in Britain, at a secret pond deep in the New Forest; almost certainly introduced, but just possibly – exquisitely possibly – native. Had Copper and Loader found that secret pond?

'Aw. Where did you get them?'

'I caught them on holiday,' said Adrian, tipping grasshoppers into the tank.

'Not that pond in the New Forest?'

'I read about that. No, I caught them in Spain. We were at a café beside a lake, and there were these bamboo canes growing up in great clumps. I think they were bamboos, leafy on top. And there was a little path going through them to the lake. We had finished our tea, and were just about to go. My dad was paying. I said, "Look, can I just go and see what might be out there?" They said "Hurry." So I ran through the canes to the lake and saw nothing. But when

I was almost back at the caff, something fell on the path at my feet. It was a tree frog. I grabbed it and started doing a dance. Then I looked up and saw loads of them up in the canes. They were hard to reach, but you could just peel them off. I got eight. Smuggled them in through the customs. My little brother was dancing around in a big sombrero hat, which made us look innocent, I suppose. I was nervous.'

'They're great,' I said.

He grinned. 'Yeah, they are.'

'Is your Grass Snake in the garden?' said Micro.

'Oh,' said Adrian.

'He escaped,' said Phil.

'Yeah,' said Adrian. 'Two weeks ago. There was a tear in the netting. I think a cat did it. I keep thinking he might turn up, down there or next door. I don't know where he would go.'

'But we're going to get another one,' said Phil. 'We're going down to Camber on Sunday on our bikes. We want Grass Snakes and a pair of Marsh Frogs. Do you want to come with us?'

'I can't,' said Micro. 'My grandad's coming.'

'I can come,' I said.

The furthest I had ridden on my bike was a park about three miles from where I lived. My father didn't like me having a bike. He was afraid I would be killed and told me to ride only on pavements. But I couldn't do that. It was embarrassing, and I couldn't keep up with my friends. Dad would never agree to a long ride like this, on major roads. I would have to say I was going somewhere near home, with Micro and these new friends. The thought of the bike ride scared me – a sixty-mile ride to the Romney Marsh coast with two older boys I hardly knew. But it was a chance I couldn't miss. Not going was unthinkable. I might have to go to that school every weekday, but here were new friends, and the wild country of southern England was opening up to me, with Grass Snakes coiled at the centre.

The bike ride was an ordeal. I kept falling behind the others, and had to stop after about twenty miles, at a village green somewhere near Tonbridge, and throw myself onto the ground panting hard. Copper and Loader were surprisingly considerate. They said that when I got my second wind I would feel better, and that they would try not to set such a pace. I wobbled off after them, and soon they were out of sight again, but they kept stopping for me to catch up, and I began to feel better. They were right about the second wind, and I felt gleeful at how much distance we had covered. I had never felt so grown up, so at large in the open world. Around me was the countryside. I could go any way I chose. When we began to see the flat fields and long straight ditches of Romney Marsh, and I realised we were nearly there, I felt flooded with pleasure.

We set out slowly along the steep bank of a drainage ditch. Ade and Phil were ahead of me, Ade near the water with the net and Phil higher up the bank. I followed closely with the rucksack and the big tin with handles we had brought. I was glad to do it. As soon as we started we heard a series of plops. 'Marsh,' said Adrian. 'There!' He pointed to the middle, where a large green frog was watching us, next to some floating weed. Just its eyes broke the surface of the water. Against the dark peaty water, the frog's body was clearly visible, one leg half-extended. Its face seemed to be grinning. I saw the thick vein along the side of the back, and the perfect circle of the eardrum, looking too large. We crouched. 'I'll try and get him,' said Ade. 'He's a bit far out.' The white net began to move through the dark water. Ade leaned out, and slowly brought it round under the frog, but he was at full stretch holding the end of the handle, and his grip was shaky. As he swooshed the net upwards beneath the frog, there was too much disturbance, and the creature shot forward, straightening its legs in the dive. That was the last we saw of it, for the net came out of the water with a splash, and slapped down on the surface. 'Damn,' said Adrian, and then, like a cowboy, 'day-am!'

As we walked, there were always splashes ahead of us, as the frogs went in, some tumbling and somersaulting yards down the bank on their way. It was terribly hard to see one before it jumped. Ade spotted one eventually, down beneath us, at the very edge, on a patch of bare mud: an unusually light-coloured yellow-green frog. They go lighter and lighter when the sun is on them. The thought occurred that we could crawl down a few feet to the side of the frog, and slide the net along in front, so that the frog would jump into it, but as we whispered there was a plop and the frog was gone.

On we walked. The ditch seemed to go on forever. We couldn't see over the bank into the fields. Hoverflies hummed. The sun was hot, and there were cracks in the mud; sometimes loose earth pattered down from Phil walking above. Light blue damselflies jinked and hovered above the water, some of them mating in mid-air. Among the splashes ahead of us were tiny plops, sometimes a sprinkling of them like rain. We realised these were baby Marsh Frogs, half an inch long, and once we'd seen one, we started to notice more of them, on the mud and on top of the blanket weed: perfect miniatures with tiny goggling eyes.

Phil launched himself forward like a rugby player diving for a try, and rolled down the bank ahead of us. 'Grass Snake!' he shouted. 'There it goes,' shouted Ade. I saw a yellow spot come racing down the slope, turning this way and that and disappearing into the grass by the water's edge. Phil had missed with his grab and tumbled down before the snake did; it now hit the water and raced in wide curves across the surface, until abruptly it stopped, letting its unfinished curve loosen and sink. The head remained above the water. It was as if the snake knew it was out of our reach. The spot I had seen was the snake's yellow collar. We watched in silence. Phil had stopped himself just short of rolling in, and now sat with his knees up to his jaw. 'Day-am,' said Adrian again.

We ate our lunch watching the snake, which didn't move.

'Still, it's good that we've seen one,' said Ade. 'We know they're here.'

As we sat there, a movement caught our attention. A baby Marsh Frog was hopping on top of the weed. Almost idly, Phil skimmed the net through the water towards it. The frog didn't seem to see, and Phil got close enough to make a wild twisting lunge which drew in a great hank of blanket weed. We thought there would be nothing in the net except weed, but the little frog was there, and another; we'd no idea where that one came from. The small ones were easy to catch once we learned to wait for them. Less cautious than the adults, they climbed out of the water soon after jumping in, and sometimes didn't dive at all, but floated with one leg outstretched. In no time we had twenty. I caught three of them myself.

'I'm not sure how to keep these,' Adrian said musingly, as we watched them in the tin. We had put in an inch of water and a lump of blanket weed. Pairs of eyes were everywhere in that water. 'We'd have to get really tiny insects. I don't know if we can. Would they eat greenfly? I don't know.'

'I could put some of them in my pond,' I said.

To my surprise, this seemed to be the answer. 'Yes,' said Phil.

'Yeah,' said Adrian. 'That's right. They don't ever go far from the water. So in a garden pond they might stick around. Yeah.

'Of course,' he added. 'They'll still be ours, partly. We'll have to share them. A joint venture.'

I hugged myself with delight. Marsh Frogs in my garden! Shared with Copper and Loader!

'When we catch big ones,' he said, 'we'll work out how to build the right enclosure. But these we'll keep in your pond. If some of them grow up, we'll dig it bigger.'

'Yes, of course.'

'Now let's get a Grass Snake,' Ade said.

We walked on, and the next Grass Snake chance fell to me. Adrian, ahead of me, must have missed it, looking for frogs in the river. Or maybe Phil, above me, disturbed the snake and sent it in my direction. I was daydreaming, looking at the ground, and the ground was moving. I couldn't focus; the movement was this way then that, and I saw the yellow spot coming towards me, and part of the snake here, and another part there, and it was all sliding fast, and I grabbed with my hand; nothing there. I grabbed again, and the movement went through my legs and was gone.

'Grass Snake!' I shouted. 'I missed it. Agh! It was right under my hand.' Such a chance. I so wanted a Grass Snake.

We saw only one more that day. Phil, at the top of the bank, had gone ahead of us, and Adrian pointed out to me a nice Marsh Frog in the middle. So this time we didn't see Phil's diving lunge. Walking on, we saw him sitting a little above us. He was smiling, a self-mocking smile. 'Come and see this.'

'Have you got one?' said Ade.

'Not exactly.'

His hand on the ground was holding something.

'A big one. It shot down the bank here and into this hole, and I

135

managed to catch the tail. But most of the snake was already in the hole. Now it's gripping so hard that I can't pull it out. I would tear it in half. I can feel its skin stretching.'

We looked at the olive green tail in his fingers, tapering to a point, and the thicker part disappearing into a crack in the bank.

'It's ridiculous,' said Phil. 'I'm actually holding it. But we can't get it.'

'Could we dig it out?' said Adrian.

'I chipped at the bank a bit. It's rock hard. And he's pulling through my fingers. I can't stop him. He'll tear himself.'

'You'll have to let go then,' said Ade.

'Can I hold him first?' I said. 'I've never held one.'

'OK. Take the tail below my fingers.'

I did. Phil loosened his fingers, and the last bit of tail sprang free from him and twined around my thumb. I felt the warmth of the snake, even in this sharp little extremity; I felt the strength too. Phil was right. The snake was pulling hard, and I could feel the body stretching. It would rupture.

'I'm going to let go.'

I was holding a Grass Snake.

I let go, and the tail went into the crack.

But there was one more excitement to come.

We were walking back, relaxed, not looking very closely any more, and I was thinking about the ride home, when Ade, who was behind us, said 'What the . . . !', and ducked down low as he went down the bank, gesturing behind him that we should duck too. We followed. 'Look at that,' said Ade.

'Bloody hell,' said Phil.

On the blanket weed, facing us, was an enormous frog, simply gigantic, four or five times bigger than any that we'd seen. It didn't look real. 'I can't believe it,' said Phil.

'Better believe it,' said Ade, and swung the net face-on towards it. The frog jumped and hit the water with a loud slap, as Ade twisted the net, swept it down after the frog, and hauled it on through the depths of the ditch, dredging up mud and weed, and out of the mouth of the net, as it came towards us, the huge frog hopped, and was on Phil's stomach, and he had it in his hands. 'Quick, the tin,' he shouted, and I had the lid off, and this animal, too big for his hands, was stuffed into the tin, and I got the lid shut.

'Oh my God,' said Ade. 'We got it.'

The fact seemed unbelievable. That extraordinary creature, out there on the weed, beyond our imagining as well as our reach; bigger than anything we had prepared for. We started giggling wildly. In the tin, there was bashing and sloshing. The tin shook. It fell over. Luckily the lid fitted stiffly.

'Those little frogs are in there. He's going to crush them.'

'Let's get him to the road. If he gets out then, we can catch him.'

So we ran. When we opened the tin on the pavement, Ade had a pullover ready for the frog to jump into. 'Get a load of blanket weed,' he shouted. 'A lot.'

I brought back a great scarf of the stuff, and after Phil and I had removed all the small frogs, which amazingly seemed unharmed, and put them into jars, we gathered round the pullover that Ade was clutching tightly to his body. 'If we put him in the tin and really stuff it with the weed,' Ade said, 'then he won't be able to jump and hurt himself. He'll be trapped and he'll be cushioned.'

So, gingerly, all of us leaning in closely, we began to unwrap.

And the frog leaped out, sailed through six groping hands, and was off down the pavement.

'He went under that car.'

'Right. You go round the other side, and you to the front.'

I got down and looked under the car. The faces of my friends appeared also. Phil was facing me directly. Ade was off to one side. The frog was in the middle. It was as wide as a dinner plate. The rubbery jawbones were like huge lips, the face more like a pike's face than a frog's. Ade, who was closest, reached an arm in. The head went down, as if in submission, the eyes squeezing shut. But then the frog jumped, and hit its head on the car's underside; falling back, it moved in Phil's direction, and he brought a hand down on it. Phil's hand was large, but came nowhere near covering this frog. He had the beast pinned, though, and my hand was on it, too. We held the frog still, and he got his other hand around it, and was able to withdraw it from the car.

Poor frog: I wish it had escaped us. I have not seen one anywhere near that size since. It must have been exceptionally old. Like toads, they can live a long time. Fifteen years has been recorded. But in the wild they rarely do, as the larger a frog gets, the more conspicuous it is. Or was this an escaped American Bullfrog? Several small colonies have been found in south-east England in the last twenty years.

We weren't home until eight. I was frightened that my parents might have been phoning around looking for me, but they were unconcerned. Mum just said that I was a bit late. My heart was thumping. I slipped out and tipped the baby Marsh Frogs into the pond. For the whole of that spring and summer, when I approached the pond, I heard splashes as they jumped in. Four were still there the following spring. But the giant frog I didn't see again. Ade took it home, and put it with his Common Frogs for the night, in a long zinc bath with a cooking bowl buried for a pond. There was net curtain over the top, held by a wooden frame. Ade put two bricks on the frame, for added weight; he thought that would be enough. He was wrong. In the morning, the bricks were on the ground, and the lid was on crooked, so that a section of the bath was uncovered. The frog, pounding against the top, had managed to dislodge it. What happened to this giant frog in the suburban streets and small back yards, we never knew. Perhaps surprisingly, no word of a huge frog came to us. Was it hit by a car? Did someone find it and take it to a lake, or call the zoo? The houses in Ade's street had small narrow gardens, not large ones with ponds like the ones in my road. There was no refuge for it there.

This spring I tried to find the ditch where we caught the giant frog: the place of my first safari with Ade and Phil, one of the best we ever had. I thought finding the spot would be easy. The ditch, I remembered, began next to a caravan site, only a road from the beach. No such place was there. I went through the long village three times.

After dark I went out to hear the Marsh Frogs call. On the road from Rye to Appledore, with my car window down, I heard them faintly, a cackling, a long rattle, in the air. A ditch ran beside the road. I parked and walked along, hearing the throaty laughter, the strange nightingale, ahead of me. Keeping on through the surge and fade of a car going past, the call grew louder, a rapid stutter, a dry chuckling, speeding up to a rat-at-tat-tat, and slowing down. Other voices sang in the distance, but one was just ahead. It was really loud now; the frog must be beside me. I switched on my torch and found him. There he was, between reeds in the water, the laughing frog, a large male, but nothing like the size of the monster we found that day. He was laughing, a muddy green frog with a purple balloon either side of his head, swelling tight and subsiding. On the Marsh it is easy to find them. Just listen by ditches at night.

Chapter Four
Common Lizard, Slow Worm, Sand Lizard

The Common Lizard, Zootoca vivipara, *is a slim brown lizard, about* two inches in adult body-length, with three or four more for the tail. They are the colours of old bracken, old grass, rotting logs and weathered fence-posts. These colours are subdued. No vivid mark stands out. But the patterns, on the tiny bead-like scales, are intricate, with flecks of dark and light, and shades of burnt wood and brown leaf. A pale line mimics a yellow grass-stalk. A dark line mimics a crack. There is depth in these patterns, suggesting the micro-landscape of the heathland or woodland floor – the grass mulching-down from last year, the fallen leaves, the log-piles and tangles of sticks, the mounds of gorse-thorns, the crumpled bracken and patches of bare brown soil.

They are to be found in every British county, up to the tip of Scotland. In woodland, heath or dune, these lizards can be seen basking on the verges of paths. Look for them especially where there is heather, bracken, gorse, loose rocks, quarries or cliffs. Meadows may have them, if there are not many pheasants. The lizards will be at the meadow's edge, where there is cover. Railway banks frequently have them. Common land in the midst of a city may do so. If your garden backs onto any sort of tangled bank or wasteland, and you keep some of it wild, you may find Common Lizards there.

See one run along a log, disappear down the other side and come up again, grinning. They love logs and stumps for basking, and the loose earth of new molehills, warm in the sun. One launches itself at something, and turns to face me with a spider in its jaws. On both sides of the lizard's mouth, legs are waving.

Sometimes they bask in small heaps together, scrambling over each other and resting a chin or tail on another's back. Adults do this, and so do the dark brown babies, whose tails darken to blackness towards the tip. Like baby toads, these youngsters seem dusted in gold. The heads of the babies are large for their bodies, the shining eyes large for the heads. In proportion to the body, a baby's tail is shorter than an adult's. All this makes their movements look clumsy, but these babies are fast.

My approach makes a group disentangle and scatter. They vanish down tunnels in the grass. Often, all you see of a lizard is that retreating tail. But after a minute, an enquiring head appears, and another, wearing what I can't help but see as a smile. Or, as I approach more cautiously, a lizard cocks her head sideways to look up at me. They often do that. Should I run, her body wonders. What is happening up there?

In Britain, a brown lizard is nearly always a Common. There are a few places where it may be something different. Are you on the Dorset heaths, or the Surrey heaths, or the dunes near Southport,

Prestatyn or Harlech, or on Coll, the inner-Hebridean island where corncrakes rasp, or at Dawlish Warren where pale dune-sand blows onto a brick-red beach? A lizard has caught your eye. It is sheltering under the quivering grass on a dune. You drop to a crouch. Lizards are most alert to movement above them, because of the danger from birds. Your lizard is still where it was – a concentration of colour and precise form in the vagueness of grass and sand. If you make the wrong move, it will be gone.

Often, I have spotted a lizard, crept up and found myself staring at an empty patch of sand. In a blink, the animal has vanished. The space where the lizard was basking is empty, but has now become a defined place, no longer lost in the random sweep across sand, grass and heather. There was a lizard here. The creature's absence dominates the place, which has become a centre, with things positioned around it. Framed by *that* broken stalk, lying at a particular angle, and *that* branch of heather, and *that* white shell, the place is an empty stage.

As I walk away, I note more landmarks. Herpetologists doing survey work, or catching lizards to move them from doomed sites, put little flags of sticky tape on plants near the spot where a lizard ran in. I will use my landmarks when I come back to try again. They will tell me I am approaching the spot and must drop low and move slowly. The animal may be out again. Usually, if a lizard has come out, it has taken a slightly different position, changing the landscape again.

But you have crept up without scaring your lizard. It is still there. You are close now, squatting on the path by the dune. Is the lizard a light sandy brown or pale grey, marked with ocelot shapes – dabs of cream ringed in black? If so, it is a female Sand Lizard, *Lacerta agilis*, much rarer than the Common. Or has it, on its face, throat and flanks, a shining mosaic of green? It is a male Sand Lizard. Slightly larger than the Commons, they are blunter in the snout and much more thickly built. Mature males are heavy-headed and bull-necked.

Dune Sand Lizards and heath Sand Lizards differ in colour. A male on the heath has a swirling green pattern, like the tufted moss in clearings between clumps of heather. Males bred in the dunes are lighter, bright lime or near-yellow. Occasionally you see an intensely green lizard with scarcely any other colours – perhaps faint brown traces on the back, or black peppering on the sides. In Britain, that is rare. Some females at Ainsdale Dunes in Lancashire have the colours of the beach: tiny pebbles of white, brown, yellow and pink, with a larger black stone here and there.

Such differences are the effect of different landscapes. In each place, the colours that made the best camouflage became the usual colours. These lizards were once much more widely spread across Britain, when heathland was much more extensive. Reliable records place them, in the last hundred years, in dune and heath sites in North and West Wales, Devon, Cornwall, Hampshire, Kent, Sussex and Berkshire. Now Dorset is their stronghold.

Daydreaming in the sand dunes there, I catch sight of a gorgeous lizard – a resplendent male, or a female whose light grey has a pinkish tinge. My heart tightens. I am all attention. Everything radiates from that spot halfway up the bank where the lizard is basking in the open, facing away from me. For a moment I don't dare to breathe. I mustn't scare it. Tense to the tips of my fingers, I drop to my knees. I can't see the lizard from here, but I know where it is. The glimpse I had is huge in my mind.

Like the Smooth Snake, the Natterjack Toad and the Great Crested Newt, the Sand Lizard has special protection. Many of the things we did as children are not now permitted. Deliberate capture is against the law. It is illegal, too, to disturb listed animals in any way that threatens their survival, reproduction or hibernation. But careful watching – crouching, kneeling, edging closer – is no threat to any of these things. Fear of the law should not deter people from going out to look for these animals. Natural England, the National Trust, the RSPB and other owners of reserves invite us to do so. Boards at reserve entrances show dramatic pictures of the different animals and say where they are likely to be seen. Wardens give advice about the best places. But one should not trouble the animals by chasing them, trying to handle them or damaging their habitat. Often the path gives a very good view. Studland Beach nature reserve is covered in paths, and many times I have seen Sand Lizards basking inches from the edge. Occasionally they bask out in the middle. In places a bank rises steeply from the path, with ledges and niches of sand or moss. The bank forms a natural showcase.

Perhaps, when I am closer and look again, the lizard will have gone. Then I will have seen it only in one radiant flash. That single image will stay in my mind, large and bright. The real lizard I may never see again. Often you don't. You come back several times, but the place is empty. This brilliant creature has gone into a fold of the world that I cannot open. But when I get close and find the animal still basking, I relax. I can settle into watching. My eyes go slowly over the markings – those ocelot shapes, or ocelli, as naturalists sometimes call them: marks like eyes. I stare into the maze of green on a flank; see the swirl clarify into tiny scales of different colours. The pink glow on the female is where reddish sandy brown on her side meets grey on her back. Down the centre of the grey runs a path of brown, and on it a line of bright ocelli – black pansies with white centres. Here and there a few more pansies line her flanks. On a heath lizard, these blotches would form denser patterns. The overall shade would be darker, the markings more complex.

This female has sloughed her skin recently. Her colours are bright, and her scales neat and smooth. I can see the edges of the large cream-coloured scales that reach under her belly. Each fits a little way over the next. I look at her head. Her eyelids are beaded. Behind the eye, where the head joins the neck, is what looks like a hole, a dark vertical oval: the eardrum, the tympanum, strangely unprotected. On its surface I see grains of sand. Behind her arm, the curve of her body has put ripples in her skin, making tiny scales turn outwards like rigid petals. A finger running along that smooth body would ruffle them. Of course, I could never really touch her like that. She would be off, skimming the loose sand in a flicker. Or, in my hand she would be hot scrabbling panic. Wild animals when you get to touch them are surprising in their heat and solidity. Their reality as something other than images is disconcerting. One sees them and hears them and smells them, and they stay wild, but touch brings to an end the separation of spheres that is necessary to our safe contemplation of wildness. Touch gives animals the dimension of space, and puts them in our world with us. It either places the animal in our power or puts us in the wild zone with the animal, physically vulnerable ourselves.

The male Sand Lizard looks like an old warrior. He has scars and blemishes. The top of his head looks weathered and scabbed. Two large scales seem to be blistered, slightly raised. Perhaps they remained attached at the last shedding. Behind an armpit hangs a bunch of purple grapes. He is carrying ticks that have pushed their way under his scales to hook into the skin. I count seven. It is a common sight. In habitats where there are deer, lizards often have ticks, and if there are trees near a heath, there are probably deer. Usually it is the tiny larva or larger nymph that attacks a lizard; these are the first two stages of the tick's life after it hatches. The adults, in the final stage, feed on large mammals. On a lizard, it is the nymph you are likely to see, and it becomes visible only when full of the reptile's blood. Before that, the tick is a spider-like creature with waving legs – or perhaps they are arms? Its body, when empty, is circular and whitish, with a plasticky texture, hard to crush between fingers. They wait at the end of a grass-blade or leaf for a host to brush past. Sensing the body heat and sweat, and the breath of the approaching animal, they ready themselves, arms waving, for they have to grasp on as the animal touches. They cannot leap or drop. A blood-filled tick is a purple globule, with a dull sheen like a grape.

Ticks do not really hang in a bunch. Each has got onto the lizard separately, at a different time, but they nearly always collect in that area next to the armpit. Perhaps it is especially warm there, or the armpit protects them from being rubbed off. I have never seen ticks on snakes or Slow Worms, which are presumably more likely to dislodge them, sliding through plants. A tick only stays on a lizard for four or five days, after which it drops off to wait through the winter before moulting into the next stage of its life. But new ticks arrive, and over time heavy infestations and the infections they sometimes bring can stress the lizard, weakening its immune system and leaving its body in poor condition for the long haul of hibernation.

I watch on, and begin to concentrate less tightly. A breeze runs across my arm and stirs the grasses above the lizard. It does not move. The sun goes in briefly and comes out. Sounds return. The dunes have a murmur, composed of breezes in the grass, winds lifting sand and letting it fall, the wing-sounds of insects, the calling of birds, cars in the distance, a passer-by shouting to a dog, the dog barking and the sea.

Is it time now to move on? I am held in place. The lizard's stillness suggests that something is about to happen. I do not want to miss it. One more look, I tell myself, one more minute. Then I will pull myself away, leaving this lizard to the hazards of the heath. I am tense with imminent movement. If eventually I do scare the lizard, making it flash out of sight, there is a sort of relief. The experience is over. But if not – if the lizard does not stir – I find it difficult to leave. Sometimes I tear myself away, but then go back after a few paces, anxious to check some detail of the animal. I want to be back in the quiet intentness of watching – the way that moment shuts out the wider world. Sometimes I go back several times. 'Goodbye,' I whisper at last to the lizard. I say it to make this the final departure. 'Farewell.'

Lizard-watching gives extreme contrasts between movement and stillness. So much time is spent walking along paths and sunny banks, looking at so many perfect places that ought to show me a lizard but do not. That patch of moss seems just right, just framed for a lizard. No lizard is there. This log is such a dramatic setting, with piles of twigs around it, providing cover in all directions. There *must* be one there. It is empty. I am walking a path that makes a figure of eight across the heath, and I keep going round, approaching that log four times. Each time, I think there will be a lizard. I can imagine one there so easily, so vividly. But there is nothing. And suddenly I see one, in an unpromising place, only inches from the path, almost in shade under brambles where someone has left an old trainer. I have passed this place four times already. Now I drop to my haunches, heart pounding against my fixed body.

The evolutionary biologist Stephen Jay Gould says in his book *Bully for Brontosaurus* that among palaeontologists there is a superstition that the best find is always made in the very last moments of the expedition. Sometimes camp has already been packed up. TV wildlife documentaries often tell that story too, or contrive it. For a week, researchers have been trying to find an animal, sitting in trees all night, checking camera traps every morning. They have to prove to the government that this animal is here. If they can, perhaps the forest will be protected. But nothing has been seen, and their time is running out, and their money. On the final evening, they see the animal. Or the last film of all from a camera trap shows one, looking up curiously. The nick of time – it is an old, old story.

A similar story gets into the minds of reptile-hunters towards the end of a visit to the heath. Let's do one more circuit. I want to go to that log just one more time. It looked so wild and promising. We can't leave without trying once more. The heath is falling quiet now, but the evening sun is hot, and its rays are going into places that have not had sun all day. Something really big might come out. I can feel that it is there. And sometimes it is. On a farewell circuit like this, I saw my largest ever male Sand Lizard, with a tail about nine inches long. He had climbed to the top of a twig pile for the last touch of the sun.

When I was doing rescue work, that final search of the day became poignant as the season wore down. On a patch of heath about to be built on, a square already surrounded by houses, I was capturing animals for relocation. For this I had a licence, obtained for me by the British Herpetological Society. The recommended way of catching lizards was to use a noose made out of fishing-line, a sliding loop that could be could be enlarged with a finger and then pulled tight. It had to be large enough for the lizard's body to pass through. Tape the noose to a stick, and sidle it up to the animal, taking care not to come from above. Slip the noose over the lizard's head, and pull it tight. The lizard's first impulse will be to lunge through the noose, so that the line tightens around the body, not the neck. You can then withdraw the stick, lifting the lizard like a fish on a line, flapping and twisting.

I remember the last Sunday of one September. A showery morning had kept the reptiles under cover. By this time of year, some adult Sand Lizards have gone down for hibernation. Others have stopped eating in preparation, and are therefore not very active. At some time in the winter, diggers and bulldozers would be arriving.

After lunch, the sky cleared, and I dashed to the heath. The warmth brought out of the damp heather a bitter smell. By four, I had a carrier-bag full of Common Lizards scrabbling frantically at the plastic, Slow Worms uncurling and lashing, and baby toads worried and blinking. But the few Sand Lizards on the site had been elusive. One I had managed to grab by throwing myself down as soon as I saw it in a heather clump next to the pavement – a technique remembered from childhood. Another, a female, I spotted on a log surrounded by brambles. This might be the last time she would be out. She was fat for hibernation, and looked placid. My arm had to go over the brambles to bring down towards her my fishing-line noose, taped to a stick. She was almost too far for me to reach. If I could slip the noose over her head, I could then whisk her up.

The method was to get the noose down to her level before moving it close. She was alert to things coming from above. Across the brambles it was very hard to do this. I got the noose touching her nose, but then snagged it on a twig, and had to ease the noose away before trying again. It wouldn't go over her head. The noose twisted into a figure eight. As I tried to fit the lower loop over her nose, a movement of my stick must have caught her eye. She turned back upon herself, paused for a moment and shot out of sight.

Three times I went back, the last time more in hope than expectation, as it was nearly six and the log was in shadow. There was no sign of the lizard. Most of the ensuing week was wet, and October began with windy rain and flurries of leaves. That warm afternoon on the heath seemed a distant land.

I kept having a dream, in which I arrived late at the new school one morning, and there was no desk for me in the classroom. I stood by the wall at one side. When the master came in, I said, 'There's no desk for me, sir.' He looked at me. 'Who are you?'

'Kerridge, sir.'

'Kerridge. You don't mean Burridge?'

'No, sir. Kerridge'

'Well, what do you want?'

'My desk isn't here, sir.'

He looked puzzled. 'Who are you?'

I didn't know what to say.

'Well, boys,' said the master. 'Can anyone tell me? Who is this? Who is *Kerridge*?'

They all turned to look at me. No face showed any recognition.

'Open your books,' said the master. The lesson began. Everyone was murmuring over some conversational task they had been given. I stood there. The master clapped his hands. Everyone fell silent. I stood there. The master asked a question. Hands went up. I stood there. On it went. And I woke with a hard ache in my stomach.

My dad confronted me over my going out every evening and not making friends at the new school.

'I've already got friends. Why should I make new ones?'

'It's a good school. We sent you there because the grammar is going to go comprehensive. It's already not good enough. We want you to do well and go to university. You're not giving this new school a chance.'

He stared at me, baffled.

155

'Please, Richard.'

'I want to be with my friends.'

'I know, Rich. I know. But not every night. You need to take more time over your homework. Those boys who went to the prep school must be ahead of you. You probably need to catch up. Stay in and work during the week. You can see Micro at weekends.'

He always pronounced the name a touch sarcastically. Perhaps this time he was trying not to, but the effort produced the same awkward emphasis. I flared up; it was my chance.

'You're such a snob. You think my friends are common. You're afraid I'm going to start eating ketchup and sliced bread and Kraft cheese slices because of them. So you won't have those foods in the house, and you don't want my friends in the house either. It's pathetic. You think you're better than Micro because his dad's gone away and his mum works at the Co-op, and his grandad paints toy animals for the toyshop.'

I was shouting. He looked puzzled. I had not told him about the toy animals, not sure it was the sort of thing that would horrify him, like the processed cheese slices.

At times he would joke about the ketchup.

'All right then, eat it,' he would say. 'Get it down you. You'll be transformed, and unable to remember the person that once you were.' He laughed and rubbed his hands together, and I laughed too.

But now he looked at me seriously.

'It's true,' he said. 'I do want better things for you, like books,

156

and music and paintings, and educated ideas. You weren't getting that before.'

All at once he looked sad. 'I had to fight for those things.'

'I don't like the school,' I shouted. 'The bloody teachers are called masters. Why can't they just be called teachers? They say words I don't understand. All the other boys know those words. I hate it, Dad. Look' – I was sobbing now, but an idea came that gave me a surge of hope – 'look, okay, I promise to read all the stuff you want. I'll go to museums with you. I won't eat ketchup or Cheese Singles. I'll do everything, if you let me go to the school where my friends are.'

This really seemed a solution.

He was silent.

'It wouldn't work,' he said at last.

'Why not? Just tell me why not.'

'Oh – it can't work like that. Working on your own – it won't give you what a school like that can give. You'll understand when you're older. This is a chance for you, Richard. A wonderful chance.'

'Well I'm not taking it! You can't make me! I won't read the bloody books. I won't call the stupid teachers masters. I won't make any friends there. I'm going to keep the friends I've got.'

And, without being sent, I ran up to my room.

Next morning, my mother said, 'You know, you can talk one way at school and another with your friends. Why don't you try? It doesn't have to be a battle between the two.'

In truth, I was not getting on well with Micro. My fear was that because he could see Ade and Phil at school every day, he would get closer to them and I would be left out. Micro did not, in fact, seem all that interested in Ade and Phil, and talked more about boys in his own class, but this did not reassure me. He might be pretending. Surely the Camber trip and the Marsh Frogs had made him jealous. I kept saying to him that the little frogs in the pond were mine and Ade's and Phil's and not his, which irritated him at first. 'Can't you shut up about them?' he said. But now he didn't seem bothered. I wasn't sure, though, and never left him alone at the pond. He might steal the frogs or hurt them. Remember the little toad, I thought. Micro didn't always seem pleased when I knocked on his door in the evening, either.

That made me all the more determined.

'Can I go round to Micro's?' I said to my father. 'I've done all my homework.'

'Not again! We agreed you would only go at weekends.'

'Agreed? We didn't *agree*. You gave orders, like you always do – even trying to control what kind of cheese I eat, and ketchup. You're a bully.'

'Don't talk to me like that,' said Dad. 'It's not about cheese slices or bloody ketchup. It's about your future.'

'I'm going out,' I said.

'You are NOT going out. That's it. I'm telling you. You can go out at weekends and one day a week in the evening. The rest of the time, stay in. Do your homework. Do some reading. Do something with some intelligence to it. That's the rule from now on.'

'You're a bully!' I shouted, tears running down my face now. 'You want me to lose my friends. They won't want me any more, and you'll be happy because you think they're common.'

I ran stumbling up to my bedroom, and slammed the door as he shouted 'Just listen!' behind me.

'No, *you* listen,' I sobbed out. 'You never do. I hate you.'

Often these shouting matches ended with me being so rude to him that he threw a slap at me, or a cuff on the side of the head – 'a clip round the ear', as everyone said in those days. After hitting me, he seized up and stood there not knowing what to say, his face red. I could see he felt awful. Often, I think, I deliberately made it happen.

This time, he came marching up after me, calling my mother to follow, and shouted at her, 'Explain to him. Explain to this boy why he has to stay in and do his homework. Explain why we want him to be cultured, and not hang around the streets and parks every evening. Explain. You explain. He won't listen to his father.'

'Listen to your dad, Rich,' she said. 'He loves you.'

'I don't want to,' I shouted. 'I want to go out with my friends. He's a snob and a bully.'

Baffled, I kept on, in a defiant mutter: 'I want to go out with my friends.'

My father now took his shoe in his hand. It must have been unlaced on his foot. I thought he was going to try to spank me with it, in some sort of formal way, like in *The Beano*. Instead, he started hitting the airing-cupboard door with it, hurling himself at the door in an action like a bowler running up to the crease – cricketer that he was. He did this once, and again, and again – three times, stopping still as if he might have calmed down, and then exploding into this action again. My mother and I watched him. And he walked out of the room.

There are two other lizards that you might see wild in Britain, though they are not native species. On the cliffs of Bournemouth or Purbeck, in the quarries of Portland, at the edge of the beach at Shoreham in Sussex, or near Ventnor on the Isle of Wight, a small brown or green lizard may well be a Wall Lizard. In central and southern Europe, this is the commonest lizard. There are many sub-species, but the Common Wall Lizard, *Podarcis muralis*, is the one you are likely to find loose in Britain. The northern Common Wall Lizards are light brown on top with dark flanks and dark brown and white mottling. Further south the males have green backs with black mottling, and black and white mosaics on their sides, head and mouth. Most of the British colonies will show you both colours.

In southern Europe, they are everywhere, the most urban of European lizards. This is the lizard you might see scuttle across the pavement or run up a wall in the centre of a town in southern France. They live in gardens, on river banks and railway banks, and in other patches of urban wildness. I have watched them on railway platforms run across to the edge and disappear. Seconds later a head pops up. In French or Italian towns, river bridges are good places to watch *Podarcis*. Lean over. See them skim across the stonework below.

Wall Lizards, and the much larger Green Lizards, were sold commonly in British pet shops until legislation in 1994 ended the trade in wild-caught animals from Europe. This was long overdue. It was a trade that killed thousands. But these creatures in pet shops were part of the landscape of my childhood. Over the years, many of these superb climbers, faster even than the native lizards, escaped the fish tanks and metal baths in which they were kept. In some places, perhaps, they escaped in sufficient numbers to breed and start colonies. But most of the populations in southern England were probably introduced deliberately, some, it seems, as recently as the 1980s and 90s. The Ventnor colony is thought to be the oldest. Rumours date it from 1841, and local folklore has it that the first animals came from an Italian ship wrecked on the rocks below the cliffs where the lizards live now. These successful colonies are in some of the hottest parts of England, natural sun-traps in the extreme south of the country where lizard eggs have a good chance of hatching. A run of bad summers can wipe out colonies even here. Recent searches in the Purbeck colony at Winspit, where I have seen Wall Lizards in the ruins of wartime coastal defences and quarrymen's cottages, have found nothing, and the population may have died out.

Look for a lizard not much larger than the Common, but more

streamlined, with a sharper head, though the nose is rounded, and a much longer tail. These lizards are much more mottled than Common and Sand Lizards. Around the throat and on the flanks they are densely speckled with chips of white and dark brown or black. Males in the breeding season have a line of blue scales along the base of each flank, glinting out. Their backs are brown with dark brown mottling, or on southern males green with black marks that are sometimes speckles, sometimes waves and sometimes strange Rorschach-like patterns. Sometimes the black looks to be the base colour, and sometimes the green. They can trickle down walls fast as water.

At Bournemouth, on one of the paths that zigzag up the cliff from the beach, you might see one of these. They peep out from cracks in the wall. Or you might, by a rare chance, be looking at a juvenile Western Green Lizard, *Lacerta bilineata*, if the animal is brown and stocky with yellow green under the jaw and a thin yellow line down each edge of the back. Green Lizards too are common across southern Europe, though all of these creatures are declining. They start about half way down France, and are native to Jersey. The adults are a splendid sight indeed: huge large-headed lustrous green lizards, whose freshly-sloughed bodies, pinpointed with gold, have an extraordinary glow – green meadow grass transformed into a jewel. They crash through the undergrowth like small rhinoceros or wild boar. Like any lizards, these animals when out of condition, just after hibernation or exhausted by fighting and breeding, can appear faded and muddy, but in their splendour they are matchless. There is not much tradition of lizards being useful, though there is this strange recipe in Topsell:

> Take seven green Lizards, and strangle them in two pound of common Oyl, therein let them soke three days, and then take them out, and afterwards use this Oyl to anoint your face every day

This, he says, was a cure for acne and baldness. I feel sorry for the lizards. They were also said sometimes to warn sleeping people of the approach of snakes by scratching gently at the sleeper's face, which is a nicer story.

Green Lizards, like Walls, were once a staple of the pet trade, and often escaped, but their eggs need hot weather for longer periods, and they are much more conspicuous targets for collectors. An introduced population on the Isle of Wight survived from 1899 to 1936, but most others have died out quickly, and the Bournemouth colony is the only one presently with us. They were noticed in 2002 by the herpetologist Chris Gleed-Owen. DNA analysis reveals these lizards to be descended from North Italian animals, no doubt imported by pet shops before the law changed. Inaccessible cliffs and one of the hottest micro-climates in the country have enabled the colony to last until now, in spite of some determined collectors, but recent cold wet summers have hit their breeding badly. See them while you can.

Most likely, though, the lizard you see will be a Common, as were the wild lizards that first delighted me: a bracken-brown, wood-brown, washed-out brown, burnished-brown Common. This slim lizard here has no large markings, but subtle dark flecks and light glints in the brown, and flashes of bright orange under chin and on belly. It is a male. Females are lighter, like very old bracken, and may have dark sides or darker lines down the edges and centre of the back. A female's underside is lovely, made of large smooth scales, almost square, in lines crossing the belly. Pale yellow, they gleam like enamel. Some are primrose-coloured. A male's belly is orange with black speckles. In early summer the female is plumper and wider, because she is carrying young. After the birth, she has a conspicuous fold of loose skin along both sides of her body, until she fattens out for winter.

Drop down and keep perfectly still. They are gone in a blink. In lizard country, especially in spring, one hears again and again the short scurrying rustle as one runs for cover. But they come out again. After a minute or two, the head emerges, cocking to one side and then to the other. Everything seems safe. There is no movement above. The lizard takes a few steps and pauses again. There is no movement still. The tongue tests the air, and the animal comes out fully, looking around for the warmest place, where it stops and shuffles its feet for a moment, before settling and flattening its body.

It is hard not to see Common Lizards as enthusiastic little creatures, quiet and modest compared with big glossy *Lacertas*, but tenacious. They run along logs and up fence-posts, bending grass with their weight, crossing paths and climbing gorse and heather. Their tongues, only slightly cloven, blunter versions of a snake's tongue, lick the edges of their mouths and taste the air, collecting knowledge. They have gentle dark eyes fringed with beads. Like the dolphin's, their face wears a permanent smile, and the constant licking of lips suggests eagerness.

On the boardwalks now used in boggy nature reserves, these lizards are a frequent sight, as they love the warm wood. Running across the board they barely seem to touch it. If they keep still, it is to bask in the sun, for which purpose they flatten their bodies, pushing out their sides, to enlarge the surface of skin receiving heat. From vertical wood or stone, they can hang by their long toes, basking like that for an hour. In the flank, behind each forearm, you will see a pulsing, quite a large tremor, as the skin moves in and out: not the heartbeat, but the working of muscles surrounding the ribs. This is how lizards bring air to the lungs. They widen their ribs to suck it in and then close them to push it out.

The Common Lizard has a wide continental range, dropping south to northern Spain, where this lizard is a mountain species, and rising to the Arctic tip of Scandinavia and northern Russia, where the Common thrives in cold no other lizard could survive. It can do so because, unlike all other European lizards save the Slow Worm (which does not live so far north), the Common Lizard gives birth to live young; hence its other name, the Viviparous Lizard. By basking, the female warms her body to develop the young. She does not need to find soil that stays warm enough to incubate buried eggs. This difference prompted herpetologists in the 1990s to investigate whether this lizard, then *Lacerta vivipara*, was so distinct genetically from other *lacertid* species as to merit its own genus. They found that it was, and the Common Lizard is now *Zootoca vivipara*, the sole species in the genus *Zootoca*.

Vivipara is found in many countries, but for me this lizard's colours evoke British landscapes, dull and bright in turns, often drying after rain. The mulch of leaf litter on heath, bog and woodland produces browns like these. Common Lizards are the yellowish and rusty browns of autumn, and the greyish brown of the leftover grass and leaves in spring. There is beechmast brown in the colours, and sometimes the sheen of sweet chestnuts. Rain makes peaty soil and leafmould a fragrant dark brown, almost black. That colour is on the lizard's flank, and so are the different shades of the soil as it dries. Fragments of stalk and leaf and crumbly rotted wood make up the litter on the surface of that soil, producing tiny specks and glints. These are also on the lizard. *Zootoca* is the brown of fence-posts, logs and drifts of leaves; boardwalks too.

They bask most in spring and autumn. Basking is needed to raise their body temperature to the thirty degrees centigrade at which they can be properly active. Lizards are cold-blooded, to use that dubious but evocative term. They are ectothermic. That is, they do not warm their bodies internally, but need sources of warmth from outside, and must regulate their body temperatures by seeking heat and then withdrawing from it. So do we, of course, but the need of a lizard or snake is more extreme. If the external temperature drops, the lizard quickly becomes sluggish and immobile. An old trick of wildlife photographers is to put reptiles in the fridge for an hour or so before the shoot. After this, the animals are conscious, open-eyed and upright in posture, but stiff and barely capable of movement. Arranged in a dramatic position in a constructed scene, they hold the pose without collapsing or moving away.

At eight or eight-thirty, with the ground still moist, the first lizards out will be easy to catch and cold to touch. I love arriving at that time, when dew hangs from the spiky heathland grass. In captivity, some lizards seem only to drink from these droplets, never going to the water dish. Perhaps that is true in the wild. Dew will sparkle on spider webs as well, in late summer or autumn. I walk softly, full of rising excitement, wondering what I will see. By eleven, on a hot day, the lizards will be hypersensitive, gone in a flash. At twelve they are down in the roots or the earth, for, just as cold slows them, heat speeds up the metabolism. If they did not go in, they would start to be panting and stressed. An overcast day that is not cold or windy may be better for seeing lizards than hot sun. In early spring, the lizards are just out of hibernation. Shabby and thin, they are sometimes still muddy from the den. They need to feed up and regain condition. The winter of suspended animation has left them depleted and weak; they need vitamins, but before they can hunt, they need to raise their temperature each morning. The nights are still cold.

By early April, unless spring is late, they have shed their skins, and the hormonal changes are at work that ready them for breeding and produce intense desire, constantly tugging at them. At first, it is the males that bask most, as they secure their territory and fight off rivals. This is when male Sand Lizards produce that intense glowing green. They fight more frequently and dangerously than the Commons. Confronting a rival, a male of either species will position himself at a right angle to the other, their noses almost touching. He humps his back, just behind the head, while pointing his nose at the ground: a curiously bull-like posture, which swells up the throat and presumably makes the lizard look more powerful. The other male does the same. This posturing may be enough to make one turn away and scurry off, chased by the victor. If not, one male flies at the other, mouth open. They roll over, each biting at the other, and sometimes both combatants fill their jaws. Now it becomes dangerous. The two lizards entwine, biting harder and knotting tighter. Malcolm Smith, author of *The British Reptiles and Amphibians*, the classic New Naturalists volume published in 1951, wrote that during these conflicts the green of a male could suddenly brighten; he had seen this himself. I have not, but I can imagine it, a green flash in the brown heather. Hard biting can induce a tail to drop. Legs are occasionally severed. Very occasionally, the wounds are mortal.

Some of this fighting is for territory. The victorious male has a good basking place that will draw passing females. They see his

168

glow from further off. Sometimes males are already following these females. For a time, in late April and May, paired lizards bask together, often touching. One rests its head on the other's back, or – in a posture that seems startlingly human – the male puts a foreleg across the female's shoulder: putting his arm around his girlfriend. It really looks like that, but does not indicate monogamy. Both males and females mate with several partners.

Mating is violent. The male starts by grabbing the female in his jaws, usually near the base of the tail. If she is unwilling, she twists and bites back fiercely, which may scare him off; sometimes he drops his tail to escape. A willing female is passive, and the male then moves his bite to her lower flank, pinching hard. Holding her with this bite, he bends his body to bring his cloaca up under hers. One of his two penises is then inserted. They are called hemipenes, but are complete penises that normally rest together under the skin at the base of the tail, just below the cloaca. In erection, the blood-red penis turns itself inside out, inverting itself to emerge from the cloaca. Shaped rather like a drill point, it has ridges and thorn-like protrusions to hold it inside the female's cloaca, since she will often move off while the male is attached, dragging him along. Copulation can last thirty minutes.

In late May and June, the females bask for longer, needing all the warmth they can get to help the eggs or embryos develop inside their bodies. On hot high summer days, lizards bask early and disappear by nine or ten, coming out in the afternoon at five or six. They are active again in late August and September, needing to bask in the colder mornings to warm themselves for hunting, as they build up fat reserves for hibernation. The babies appear in August and September.

To bask, the lizard finds a little sun-trap where it will not be far from cover – a clearing in the heather, a sandy bank, a low branch, a log, or the top of a sprig of heather. Piles of sticks, logs or rocks, or heaps of rubble, are excellent, for at any sign of danger the lizard has only to slip down into the pile. A path beside a sun-facing bank is ideal for seeing lizards. I always feel excited in such a place, as my eye runs from clearing to clearing. A basking place may have the shape of an arena or stage, with the lizard as the centrepiece. Sometimes the place is a ledge like a shelf, with a lizard positioned as if for display. Some lizards bask on mossy cushions that hold the lizard like a brooch or jewel. A bright male Sand Lizard can look especially brooch-like, his body curved in an s-shape and his tail by his side. In woods or shaggy heather, the path is the only surface open to the sun. Lizards will then venture out at the edges, or bask in the middle, if it is a quiet path, not often taken.

I love the thought of that, a path where I am the first person to venture for days. It is sandy, overhung on both sides. I push through. It is a hot day, silent except for the insects. Ahead of me is a clear patch of path, and there, in the middle, is a large snake or lizard, unconcerned, for there is so little disturbance here. I drop to my knees to watch it, and the creature doesn't move. Where am I? The place is forgotten, lost in a fold. It makes me think of a story that fascinated me once, and scared me too.

There was a Canadian valley, deeper in the forested mountains than white men had ever ventured, and forbidden to the Indians by the spirits. If you do wander into this place, the animals will not fear you, for you will have only hazy, distant memories of your identity and what you are meant to do. You see herds moving across the valley, great stags and wild cattle, larger than any you have previously known. Bears and lynxes stop and watch you. Inside your body, you feel a falling sensation. A memory is shouting at you somewhere, and hammering, but the sound is almost inaudible: a tiny, puzzling sound. It is fading. Soon, all you will have is the present: no time, past or future. But something changes. Something calls you back. And you get up and run out of that valley. Later you feel you escaped at the last possible moment. The story was 'The Valley of the Beasts' by Algernon Blackwood, and I found it in an anthology of animal stories from the local library, a big green book, which I borrowed time after time.

Dad wanted us all to go for a family walk. We had hardly spoken since the big row, he and I. At the bottom of the stairs he would stand looking at me, as if struggling to say something, and I would look away and walk past him, up to my room, up to my homework, or downstairs to a meal at my mother's call. In my bedroom I watched the Smooth Newts in the tank, parting the weed-forest with their hands and gliding through.

Usually I enjoyed family walks. Sometimes we found new ponds. I could lift up old doors in the nettles or, best of all, sheets of corrugated iron, and find Slow Worms and toads. The grass that had been under the sheet of iron was yellow and flattened, and tunnels made by voles were exposed. I saw a mother carrying a baby, down the tunnel and off into the forest beyond the flattened square. Under a ball of grass in the centre there was a movement. Four more babies were there, pink and helpless. I was afraid to put the iron down. It might squash them. The mother reappeared, black eyes shining. One by one, she returned for them, venturing back into the danger. 'Put the sheet down,' said my mother, after the second baby had been rescued. 'I'm afraid to,' I replied.

I liked seeing cattle, nearby and in the distance, scattered in this field and that, throughout the valley. A cow calling loudly, neck stretched out, was answered faintly. Especially, I loved to see a bull. Cows were milling together in a field, and I would look for the back that was higher than the others, the hump of the heavy shoulder. Sometimes a bull looked over a gate at us: great block of a head, with the mountainous body behind. I made everyone stop while I stared. It was a black and white bull, a big Friesian. I could feel his warm breath. His skin twitched. Close-up, the black colour was dusty. There were matted white curls on the forehead, slightly yellowed. A tear had run down from an eye.

'Come on, Rich,' said my dad. 'Let's keep going.'

'Look at the ring,' I said. The brass ring glittered. The bull raised his head, making the ring rest on his muzzle. He dropped his head, to look at us again. The ring hung clear. On the pink nose were droplets and hairs.

'Do you think I could touch it?'

'Don't be silly,' said Dad. 'Keep well back. He could swing his head and catch you. With one toss you would be up in the air, flung over his back. He would turn round and get you.'

'What would you do?' I said.

'I couldn't do anything. It would all be too quick. Well. I would do what I could. I would climb in and try and lure him away, I suppose. But I don't think it would work.'

A sudden sneeze came from the bull.

'Let's go,' Dad said. 'He's getting restless.'

We moved off, suddenly scared to be so close. When I looked back, the bull was in the same place, watching us: neck arched, brow forward, nose down. Once, we looked down at one from a ridge above the field. This bull was light brown with white patches. In the dip of his back was a swarm of flies moving like liquid.

Often the paths led through farmyards full of mud and cow splashes. When there was a man working there, I ran up to him, asking did they have a bull and could I see it. 'I'm sorry,' my dad said, coming up behind me. 'Sorry to interrupt you.' 'It's all right,' they sometimes said. Once I was taken into a dark barn with a concrete wall inside it, above which was just visible the black and white back of a bull. I heard him rumble, his hooves shifting and, again, that abrupt release of breath. His white patches shone in the gloom. Keeping me well away from the head, the man let me climb on a bucket and reach out to touch the bull's back. I was astonished at how tight it was, how warm.

I was dreaming my bull dream a lot at this time. Edward O. Wilson, the biologist, says that snakes are the creatures people dream about most. Surprisingly, I have no memory of dreaming about snakes. But I dreamt repeatedly of these bulls, the big black and white ones. In a field there was a huge bull, walking towards us, a long-legged bull; there was no fence between. We climbed over a gate and walked on, but the bull was in front of us again. Somehow it had got through the hedge – crashed through, or there was a gap. We ran to the next gate, and slipped through and shut it, but in front of us the bull was there again, black and white against the green. It came towards us. We ran to a tree, and climbed up, all five of us, but I couldn't hold on; I was sliding down the branch, and the bull paced angrily beneath me.

A family walk was not a bad idea. I didn't mind going. But, once we were walking, Dad started to organise a game. We were to walk side by side, and take it in turns to sing 'I am the music man, I come from down your way, and I can play-ay', to which the response was

'What can you play-ay?' The music man would then name an instrument, and we would all mime a performance, rippling our fingers through the air for the piano, and doing arm movements for the trombone, as we made the appropriate sounds.

I couldn't bear to do this. Perhaps my sisters liked it. I don't know. No one else was watching, but I found the game horribly embarrassing and refused to take part.

'No, no.' I tried the comic outrage that children use with their parents now – the raised eyebrow and sorrowing expression. 'Please don't do this to me.'

But this was the 1960s. They carried on. I became agitated. It was no longer a joke. I imagined my friends seeing this.

'Please stop. It's disgusting. Why can't you be normal?'

Dad should have laughed at this, but it enraged him. 'Join in with us for once, can't you? Be part of this family.'

'It looks so stupid.'

He said to my mother, 'I've had it with him! I've just had it!'

'Get away!' he said to me. 'Go on! If you can't join in and have fun with your family, just get out of my sight.'

I didn't go. I wanted to be with them, and stood hesitating.

'Are you going to play with us?'

'I don't want to look such an idiot.'

He threw his hands up in the air.

'Idiots, are we? Get away, then. Go and look for your bloody creatures, if you prefer them to your family.'

I stood there.

'Go on.'

He picked up a flint from the path, and threatened me with a throwing action. I didn't move. It landed near my feet. Mum looked at him, worried.

'Get away, I said. Go and look for bloody reptiles. We'll see you at the car.'

Another flint bounced on the path and went close to my leg. I ran off, through the gate into the next field, and for the rest of the walk followed them like a jackal, keeping my distance. Ann and Cathy waved to me. My father looked away. The game had petered out now. I had spoiled it. Mum waved too, and at the end of the walk she came over to me and said, 'Get in the car. Don't start the argument again.'

We drove home in silence. I think I knew he wanted to say sorry.

That evening, when I came out of my bedroom to go down for dinner, he was waiting again at the foot of the stairs. As I reached him, he held out something.

'Would you like an apple, Richie?'

'Thanks,' I said. 'Thanks. Yes I would.'

I took it. He looked at me. I remember him holding out that apple; the shape of his hand, soft veins.

'I thought, maybe, next week, I could take you and your friends – maybe Micro, Adrian and Phil, out to Hayes Common to look for lizards. You said you were wanting to go there.'

'OK,' I said. 'I'll see if they can come.'

In bed that night, through the wall, I heard him moaning. He often did. There was an accelerating series of sounds, rising to a shout and a gasp. Then there was muttering. I heard my mother soothe him: 'There, love, there.' For a time I thought these sounds were sex cries, but once, after they had been loud several times in the night, my mother said next morning, 'I expect you heard your father last night. He was dreaming they were shooting at him again.'

My father had been a tank gunner in France, in the weeks after D-Day. He was twenty-one. His leg had been wounded by shrapnel. One summer he had taken us to Normandy. The invasion beaches had museums. Photographs showed troops wading ashore, wounded soldiers in stretchers, tanks in the muddy lanes, earth showering as shells hit the ground. Real tanks were on display, and German beach defences. Tricolours, Stars and Stripes and Union Jacks were everywhere, and D-Day bookshops. There were racks of postcards. All this was interesting to me. I knew D-Day from my comics, though the Battle of Britain fascinated me more. What Dad wanted, though, was to hunt down exact places, corners of fields. A name on a signpost caught his attention. Bully-les-Mines. 'We were pinned down for three days there.'

I never knew details of how he was wounded. My sister Ann said recently that something had been troubling her about this. 'He told me he was wounded during a brew,' she said. 'That's all. He didn't say any more. And all this time, I thought he meant that a shell landed while they were making a pot of tea. I had that image in my

177

mind, of Dad and the other men sitting on steps in a French town square, drinking tea. His friend was killed next to him, and he got shrapnel in his leg. But now I wonder. They used the word "brew" for a battle, didn't they? Perhaps that's what he meant. I've had the wrong picture all this time.'

The village was ordinary: quiet, a church, a small café, men sitting outside. Beyond the village was a crossroads Dad thought he remembered: grassy verges, high hedges, wooden gate. It seemed featureless to me. I went with him into the field. In the corner was a rusty piece of farm machinery. Dad looked around. There were cowpats with flies running on them. He stood still. I spotted a ragged sheet of metal in the long grass near the farm machine, and went over to lift it, hoping for a European Fire Salamander, shiny black and blotched with yellow, to me an almost mythical creature. In the yellow grass under the metal there was nothing, but I saw a movement on the gate-post. A brown Wall Lizard ran down it. Another scurried into the grass. I waited, and the smiling head emerged from a crack in the concrete, the base of the fencepost. Around its lips there were black and white markings like checkers. Dad was standing in the middle of the field.

'I think this is the place,' he said, but he did not seem sure.

The lizard, a male, ran halfway up the post and stopped, his tail quivering. A fly had settled a little further up. The lizard stared at it, head bobbing slightly, and them lunged, but the fly was gone. I half-watched this, and half-watched my father turning his head this way and that.

Micro said he couldn't come to Hayes Common. He was slipping away from me. I felt sorry, and stayed at the phone several minutes after the call, wanting to ring him back and say something but not sure what. Adrian and Phil were happy to come. I would have them to myself. It would be like the trip to Camber. I had to tell them, though, not to mention Camber to my dad.

Ade and Phil already had Common Lizards, but would advise me in choosing a pair. We got to the heath at about eleven. It was a hot June day. Immediately, as we walked from the car to the edge of the woodland, there was a scurrying sound in the grass – the rustling dash I would come to know well. Adrian strode forward, stopped and threw himself onto the ground.

'Missed it.'

The best place was the edge of the woodland, just yards from the road, Adrian said. There were gorse bushes, brambles and deep clumps of bracken. New green ferns, coiled and shiny, had forced their way up through the papery brown mulch. We raced off and left my dad reading his paper. I was going to see my first wild lizards, apart from the one at the Bunnyhole, which I now knew to have been a Common Lizard male.

Adrian held up his finger to quiet us. He signed to us to crouch. We crawled up like commandos and followed his finger. On a log at the edge, with the bracken behind it, four lizards were basking. A large female was the colour of a digestive biscuit. Two smaller females were darker. A male with a fine long tail had his chin on the back of one of these two. As we watched, he moved off. He was restless, this male, walking along the log, disappearing behind it, and reappearing on the ground beside the log, before climbing back up to the females, who were passive. 'Go on, Richard,' said Ade. 'You take him.'

On top of the log now, his back was towards me. My chance, I thought; he was occupied with some lizardy business to do with those females, and wasn't attending to me. I went down on my knees, and moved forward, bringing my hand slowly towards him, ready to snatch and close my fingers round his body. But I was hesitant. My hand hovered. The male saw the movement and raced the full length of the log, leaping off at the end. All the females had vanished.

'Bad luck,' said Phil. 'It was a difficult place.'

I wished I had gone for the biscuity female. In my mind she was such a rich colour.

'Here's another,' said Ade. 'Over here.'

This was another big female, a shiny one who had shed her skin. Gravid females at this time of year are seeking warmth for the development of the young inside their bodies. They often bask half under cover. This one was in a mop of grass, her head visible at the top. When a lizard sloughs, its skin tears and rubs off in white tatters that hang from the body. Before sloughing, the lizard looks drab. When the old skin is gone, for a few days the animal gleams. This female's skin had an oily freshness. Her markings were beautifully clear. 'Don't try to pick her up,' Ade said. 'Bring your hand down on her, flat. Pin her to the ground. You'll feel her struggle under your hand, and then you can close it. She's on soft grass. You won't squash her.'

I did exactly what he said. The flat of my hand dropped upon her, pressing gently. She couldn't run. It worked. She was wriggling under my hand. My fingers found her. I disentangled her from the grass. Now I lifted her up. Between my thumb and the soft folds of my hand, her head appeared. She was squirming out. I held her tighter. My other hand was ready. As if I had done it a hundred times, I relaxed my hold enough for her to push herself half out. My waiting fingers took her. She was between my finger and thumb – but without her tail.

Between her hind-legs was a stump, moist with blood. When I opened the hand that had pinned her, the tail was on my palm, lashing and coiling. It twisted itself upright and fell over. My eye was on that, not the lizard. A sharp pinch in my other hand startled me, and my fingers flinched. The lizard dropped. I glimpsed her fall. The tail had done its work.

Ade laughed. 'Did she give you a nip? They do, but they can't draw blood.'

'It surprised me.'

'Yeah, it does.'

I held up the tail, the stupid, lifeless thing, mechanically curling one way and then the other. A lizard's tail has scarcely any flesh. It is a series of vertebrae, each with tiny wings of bone. Below each set of wings, the tailbone is already fractured. That is where it can snap. A muscle contraction breaks the tail here, in response to a grip, or sometimes when the lizard is merely shocked. If you look closely at a tail, you see that the scales grow in rings around it. Each ring overlaps the smaller one below, as the tail narrows. The tail severs at the base of a ring, which falls out of a larger one. It drops off cleanly. The muscle contraction that severs the tail produces spasms that last several minutes, making the dropped piece twist and thrash so strongly that it can move about on the ground. If the lizard is lucky, this dramatic movement draws the eye of the predator, which turns and seizes the thrashing tail, not the escaping body. What the predator gets is just bone, skin and scale.

In my hand, the lizard must have wriggled its body free but found its tail trapped. The contraction occurred. Males sometimes lose their tails in territorial battles, or fights over females, or to the bite of a female unwilling to mate. But mostly tails are lost in escapes from predation. There is some regeneration. In time, a shorter tail grows, spiky and plain brown or grey, but it cannot be dropped a second time. The only way the lizard can use this defence again is by dropping the tail above where the first break occurred. Often you see Common Lizards in the wild with new stumps or little dark cones beginning to grow; especially in places where there are lots of cats. Sometimes a tail is half severed, and a new one sprouts from the gap, producing a fork-tailed lizard.

You escape but leave part of yourself behind, lost forever. It is an evocative idea. When people tear themselves free of an attachment, they sometimes say that they have left part of themselves behind. Where is that part now? For people, it must still be inside them somewhere: a set of memories and feelings about someone – still there, even if never visited.

It was getting too hot. I saw several more lizards, but couldn't take them by surprise. They looked up at me. As soon as I moved nearer, they were gone. 'We might have to leave it now,' said Ade. 'Your dad wanted to be home by one-thirty. The lizards are all going in now, anyway, and they're too fast to catch in this heat. We can bike out another time.'

'All right,' I said. We returned to where we had parted from my father. He had put down his newspaper, and was standing on the path, staring down at something.

'Dad,' I called quietly.

'Don't make a noise,' he said. 'Approach from behind me, slowly.'

'You go,' said Phil. 'We'll wait here.'

'Now follow my finger,' said Dad, as I reached him. 'There. Do you see?'

It was a female Common Lizard, light brown with few markings; a small one, very plump, her sides lumpy with young. I am sure I saw movement inside her. She had climbed up a branch on a silver birch bough in the grass. Most of the bough was in shade. She had hauled herself up there to catch a shaft of sunlight, slanting through the trees. Hoverflies hung in the air.

'A pregnant female,' I whispered.

'I thought so,' said Dad. 'I've been watching for ages. Once I moved, and she ran in, but I stood still, and after a few minutes her head poked out. She seemed to be looking up at me, and trying to decide what to do. I didn't move, and she came out a bit further, put her head on one side, and looked at me again. I still didn't move. She came all the way out, and climbed up to the sunlight. At first she patted the tree with her front feet. Then she let herself relax. She's heavily pregnant isn't she?'

'Yes,' I said. 'They can have ten babies or more.'

'Do you want to try and catch her?'

'I do,' I said. But I looked at that tightly plump body. Could I bring my flat hand down on that? I thought of the severed tail, writhing and bloody; the sad stump.

'She's too pregnant,' I said. 'I can't risk handling her. It might squash her.'

'I'm glad you said that. I hoped you would.'

'I still want some lizards, though.'

'You'll get some,' he said. 'Your friends will help you.'

They did, and by the end of the summer I had three pairs of Common Lizards, in a long galvanised iron bath in the garden, with a lid like the ones I had seen at Ade's house. I made a wooden frame for the piece of net curtain, and stuck foam rubber to the base, where the frame would sit on the bath rim. Every day I emptied in a jarful of the little wolf spiders that ran from my feet in the garden, or grasshoppers gathered with Ade and Phil in the nearby allotments. One of the female lizards had given birth to seven charcoal-coloured babies, which I reluctantly caught and took back to the common, since I thought it would be difficult to get small insects for them.

In the same enclosure, I had a pair of Slow Worms, the male being the largest I had ever found, a scarred old monster who looked like a badly-carved gargoyle. Most Slow Worms have no sort of bulge at the head. The smooth tubular bodies simply end in soft eyes with delicate eyelids, a mouth and a rounded snub nose. About the mouth and under the eyes there are delicate patterns, tiny curls, dots and splashes of brown on smooth white. Big old males are different. Their cheeks fill out, and their heads become bulbous. Fighting over females leaves their heads scarred with dents, scrapes and gouges. It looks as if a dragon's head, or a dinosaur's, has been fixed to the end of a thick pipe.

My male had a rare feature in Slow Worms, one we had read about and were always seeking. His grey body was sprinkled with glints of sky blue, tiny shards. When he had sloughed, they were bright. Edward Topsell says of the Slow Worm that 'the colour is a pale blew, or sky-colour', and though it is hard to connect this description with the dark bronze adults and pale gold babies one usually finds, when I think of this male the description makes sense, though I doubt that Topsell had in mind such a scratched old bruiser. It was this living gargoyle, I would guess – as well as the little Marsh Frogs, and my Sand Lizards – that prompted Phil and Ade to suggest that we merge our two collections and regard them as shared. Next day, I went to the new school with a swagger, for the very first time.

I had two Sand Lizards, briefly.

Every colour of the heath floor is somewhere on the spring male Sand Lizard – the mosses, the heather, the brown and green grasses, the dark peat, the white stones and seashells, the sand. For the incubation of their eggs, British Sand Lizards need banks of sand that get very warm in the sun. This is true of all the northern populations – in Britain, north Germany, Denmark, south Sweden, the Baltics and Russia. Further south, warmer climates make different habitats possible for this species – grassland and woodland verges as well as heath and dune. The Sand Lizard's range extends south to Mediterranean France and Yugoslavia. Meadow habitat sometimes produces pure green males, hard to tell from Green Lizards.

In Britain, the lizard's position is precarious. Huge losses of heathland to urban expansion, agricultural changes and quarrying have reduced the available habitat to the Dorset strongholds and a few small enclaves. Small colonies become cut off, as roads and houses are built around them. Genetically, they become less diverse. After disasters such as heath fires, these small populations cannot be replenished naturally. Where there is recreational use, fires are common. Before the Wildlife and Countryside Act of 1981, collection for the pet trade was another serious threat, probably responsible for the loss of several local populations. Without intervention it is likely that the Lancashire Sand Lizards would by now be no more; perhaps the Surrey lizards also.

We had a holiday in Swanage every August. After catching those Common Lizards in London, I started reading much more about lizards, and discovered that the best place in the country, so several books said, was Studland Heath, about five miles from Swanage. All the six British reptiles were there. For years we had been coming, and I hadn't known. I now visit the heath several times every spring, and each year I find the Sand Lizards in a different place. Sometimes they bask in large numbers in the low heather, just off the beach, between the first line of dunes and the second. The forest of heather has cliffs, ravines and mossy clearings. Sandy paths are overhung with springy branches. It is all at your feet; nothing higher than your knees. Sometimes the lizards are on the high dunes, in the windblown marram grass, bitter on the tongue. On slopes of loose sand, you see marks where they have scurried. Try to climb it and the sand buries your feet. Next year, there seem to be none on the dunes. I find them inland, around the rubble of Second World War fortifications. German landings were possible here. There are chunks of concrete, with stones embedded. Rusty iron bars wiggle out of it, looking like Slow Worms. Lizards run on those bars. Tails disappear into gaps in the concrete. Heat haze trembles over heather, gorse and rubble.

I insisted that we all went to Studland. Leaving the others on the beach, I wandered into the dunes and heather. I was excited about looking for reptiles – the rare Smooth Snake was here, though I never saw one, and adders, which I did sometimes see. But, at the age of thirteen, wandering in those dunes, I also felt that the place was very sexy. The sand was warm. On the beach, sandy flesh was all around me; sand drying on shoulders and thighs. The dunes were full of secret valleys, close to the beach but hidden from it. Often I found the remains of campfires. Several times, coming over the crest of the sand dune, I found couples lying in these valleys.

Even on a crowded beach there were excitements. At Swanage the day before, I had been getting changed in the beach hut we had hired. The door had upper and lower sections; the lower was closed. Across this lower door, I looked out at the sea: the shouting paddlers, splashing swimmers, distant ships. My eye dropped idly to the sand beside our deckchairs and then fastened on what I saw, as a flush went up my body. A young man and woman on a towel were snogging hard. They were about eighteen. I was positioned behind their heads. Another towel covered them partly, but the vigour of their movements made it slide off. Bending slightly, I found exactly the right angle to see the man's hand slide under her bikini top, lifting its stiff cone. That pink mark was part of her nipple. I saw it again. And again. She had flowing dark hair, streaming out on the towel. Quietly, I pulled the upper door almost shut. Now no one could see I was staring.

My throat was tight, my heart thumping.

'Do you want an egg sandwich, Rich?' called my mother.

'In a minute,' I answered.

Now someone was blocking my view. I could only see legs. On a deckchair just feet from the couple, my granddad bit into a ham roll. My dad was behind his copy of *The Guardian*.

At Studland, the next morning, as I floundered up a dune, I was thinking of that couple, and wondering if something similar would come into view at the top. But the hollow on the other side was empty. The slipping sand carried me down, and I saw a Sand Lizard.

It was a green male, in a clear patch on the heather. He caught my eye as I steadied myself at the bottom. Everything was still. I could see him between two clumps of heather. Only one position gave me this view. It was like a peephole. From every other direction he was hidden. I had lurched into exactly the right viewpoint. Otherwise, I would surely have passed him unaware. There must be many that I miss, every time.

I tried another angle, came back, and the patch was now empty. No matter. My eye was in. Now I saw lizards all around me: fifteen or so in the next fifteen minutes. It is often like that. Your eye suddenly finds the right range. Bill Oddie, on one of his wildlife programmes, said he had been coming to Studland all his life to watch birds but had never seen a reptile. I have seen hundreds there. No doubt I miss most of the birds that Bill sees.

Sand Lizards are slower than Common. I didn't know this at the time. I wanted one so much that I could scarcely make the grab. Repeatedly, I agonised, hesitated, missed. That day I caught nothing.

I trailed back over the sand and babbled to my parents about the lizards I had seen. 'Can we come again tomorrow, Dad, please? I must get one.' But the next day was cloudy and wet. The day after that we went to the Isle of Wight, on the ferry. It was Friday morning when we went again to Studland. This was my last chance.

Desperation made me more reckless. I missed many, but it was prime basking weather, and the flat hand technique eventually caught me one that I thought was a female: it had dark flanks and a pattern of ocelot markings, with no green. Looking back now, I'm not sure. The green on a male often fades by late August, and my memory, a strong one, of this lizard's large blunt head suggests a male. With that one in my jar I walked back in triumph, and just as I got to the car park, a startled lizard ran across the path at my feet. I threw myself down, and felt it struggling in my fingers: an unusual one, a grey Sand Lizard with light brown lines down its back, and white flecks on the brown. A dark little spike was all this one had for a tail.

At home, I put them in with the Common Lizards, and the grey one, which I thought was male, disappeared for a week. I was beginning to think it had escaped, though I couldn't see how. The others were active, rushing for cover at my approach. Then the grey reappeared, with that tuck in its side I had seen in Common females after birth. It had buried a clutch in the enclosure. I didn't know where.

At the beginning of October, I made them a hibernation chamber, a *hibernaculum*, as the books called it. Having caught all the lizards and put them in a fish tank, I dug a pit in the bath, and patted the soil flat on the bottom. Into the pit, I put half a large flowerpot, on its side. Dried grass went in for bedding. I heaped earth over the lot, and pressed it down, forming a mound, leaving a narrow tunnel open for the entrance. One by one, I put the lizards in the tunnel, nose first. Each one I looked at hard, stroking its head and wishing it luck. They slipped out of sight, but one of the Commons re-emerged and made a break for it, running across the enclosure. I caught him and put him back in, placing a large tile over the hole.

Next day two of the Commons were out. They had dug their way free. I put them back in. A day later, the grey Sand Lizard was out, and I put her back too. Anxiously, I removed all the stones, logs and tussocks, so that the lizards had no shelter but the hole, unless they burrowed. The weather turned cold after that, and I saw them no more. For insulation, I piled leaves upon the mound, and heaped more around the outside of the bath.

The winter was mild. We had hardly any snow. After the first of March, I took the top off the bath every day, to look for the lizards. There was a day of hot sunshine. Surely they should be emerging. Near the end of the month I went to the common with Ade and Phil, and we saw wild lizards. What had happened to mine? April came, and at last I made myself take out the leaves and with trepidation lifted the tile, broke the tunnel and opened the chamber.

Four were alive – the Sand Lizard I had first thought was a female, and three of the six Common Lizards. They looked thin and drab and were sleepy. The others were shrunken corpses. It was damp in the hole, and the bodies were decaying. There was a smell.

The next morning, two more of the Commons were dead, and the others were barely moving, their eyes half open. I took the two survivors indoors, into a glass aquarium hastily arranged with sand and bark, and warmed with a forty watt bulb. The last of the Commons died that night. But the Sand Lizard began to show signs of life. It dragged itself to the water and lapped with its tongue. I saw that its eyes were fully open now. A few days later, it was lively. The mealworms I put in disappeared, and I saw the lizard eating at the dish.

That lizard lived three more years in the indoor vivarium, with some Wall Lizards and an African skink from a pet shop in London. The Sand Lizard ate readily. It basked under the light bulb and shed its skin. But its bright spring colours never came.

In all but its core Dorset population, the Sand Lizard in Britain now is not a product of the forces of wild nature. We have it because human enthusiasts have put it there. A work of art, as well as a work of nature, it is there because some people find it pleasing.

Surveys in the 1980s found that populations were plummeting. The New Forest Sand Lizards were gone, their last known colony wiped out by pet trade collectors. Lancashire was down to fewer than two hundred. In Surrey they were barely holding on. One of the best Surrey sites, Frensham Ponds, had lost nearly all its lizards, probably caught and sold to pet shops, says Chris Davis, one of the leaders of the Sand Lizard Recovery Programme. 'I remember biking to Frensham as a kid,' he says. 'And there it was – an animal that wouldn't be out of place in the Amazon jungle.'

Keith Corbett, a lifelong campaigner for British reptiles, produced a thorough report for English Nature in 1994, giving hard survey evidence and an analysis of every site – the sites with current populations, empty sites where the species was known to have lived in the twentieth century, and sites where there is some suggestion it might once have lived (a mention somewhere, but no verification). All the empty sites were assessed for their viability. Would a reintroduction in these places be likely to work? Corbett persuaded English Nature that there was firm evidence that a systematic recovery programme stood a good chance of saving the Sand Lizard in Lancashire and Surrey, and re-establishing it in North Wales. His report became the basis of the Recovery Programme, carried out by the British Herpetological Society, under contract to English Nature and now Natural England. This programme is one of the most significant interventions to support an endangered species in the history of British conservation. Chris Davis calls Corbett 'Mister Sand Lizard'.

Wild-caught lizards from Dorset sites scheduled for development had already been used to repopulate Surrey heaths that had lost their populations. This was the destination of those lizards I noosed that September. But one of the rules was that southern Sand Lizards should not be mixed with the genetically slightly different northern populations, and it was the northern lizard that was in critical danger – the Merseysiders, as Chris calls them. He was the first to use captive-breeding to reintroduce Merseysiders in sites where they had a good chance. Under licence, he captured eight animals, two males and six females, and kept them in outdoor enclosures, removing their eggs for indoor incubation. If possible, it was best to take the eggs immediately after laying, for then they did not have to be kept at exactly the angle at which they had rested in the sand. At a steady temperature of 23°–28°C, unlikely in the wild, the eggs hatched in much greater numbers. To prepare them for wild hibernation, the young were then fed mealworms and small crickets dusted with vitamin powder. They were released in late August and September.

In each approved site, Davis releases 50 hatchlings each year for three years. After that, 'the colony has to sink or swim.' The new population is left alone for three more years and then regularly checked. Davis does not believe in the topping-up of existing populations, except in special circumstances, such as the massive loss of animals to collectors that decimated Frensham. In most places, the losses are due to habitat damage, and if the habitat is restored, the numbers will increase naturally. If not, a restored population will not survive.

Eighty restored populations have been created since Corbett's report. Licensed captive-breeders supply the hatchlings. In North

Wales and South Devon, Sand Lizards are back, and apparently well established. From that low point in the late 1980s, the Lancashire population has doubled. Davis tells the story of how, when he began releasing lizards in dunes near Harlech, he pointed out to a volunteer an ideal incubating-spot, on a south-facing sandy slope, sheltered by marram: 'If this works, that is where we should see the first newborns.' Three years later they went to that spot in late August. Approaching, he held up his hand as a caution. Then he dared look, and in exactly the place he had hoped, a hatchling Sand Lizard was basking, near the mounds it had thrown up digging its way to the surface. 'I had tears running down both cheeks,' says Davis. 'That was one of the first Sand Lizards to be born in Wales for more than fifty years.'

Distinctions between the natural and the artificial become harder and harder to maintain. Such is wild nature in the Anthropocene. Such is its vitality.

Ashamed, I told Chris Davis about my poor lizards. 'We all did things like that,' he said. 'It was very sad for those lizards. But I don't think collecting by children has ever really damaged populations. Compared to natural losses, it is nothing. What did sometimes wipe them out was professional collecting, to sell to the pet trade.' Corbett's report tells how one of the captive-breeding enclosures was raided by such collectors; the lizards, it seems, were probably smuggled to the USA. 'I wouldn't fear for the species because of children catching a few and not knowing how to keep them,' Davis says. 'My fear now is that kids have no knowledge of them at all.'

But when I think of my two lizards, I remember the moment I slipped them down that tunnel. I think of that moment of leaving the light. In the wild, only about half the young lizards survive their first winter. Hibernation is a disturbing paradox. The creature protects its life by not experiencing a large part of its life. Topsell explains hibernation beautifully, with a hint that the process is a small miracle, a natural resurrection:

When they perceiveth that winter approacheth, they finde out their resting places, wherein they lie half dead four months together, until the Spring sun again communicating her heat to all Creatures reviveth and (as it were) raiseth them up from death to life

Some adult Sand Lizards go back down in July. They are awake only three or four months of each year.

Of course, the lizard does not choose – though the mechanism that makes some of them go down with the weather still warm is not understood. Reptile hibernation – or *brumation*, the precise term for the process in cold-blooded creatures – is a physical response. The body cannot function below certain temperatures. As autumn gets colder, the animal's metabolism slows. Lizards and snakes become less active, losing their appetite for food. For brumation they need reserves of fat, though their metabolic activity – their chemical processing of nutrients – will be minimal. There will be no replenishment until the spring. But the fat must be stored in the weeks of late summer, not the days before the animals go down. Undigested food must not remain in the stomach when the digestive process ceases. That can be fatal, especially to snakes, which digest large meals slowly. Reptile bodies seek warmth, and as autumn comes on, the only way to avoid sharp cold is to find insulated refuge, underground or in some other shelter. In these places, cold but just above freezing, the reptiles can be inactive but alive.

While they are in this state, many bad things can happen Perhaps they have not burrowed deep enough, and the frost will reach them. Shrews, rats or moles may find them and eat them underground. The reptiles will not know that this is happening. If they did know, they could not flee. Is the chamber well drained, or is there a risk of it filling with water? Will warm weather early in spring wake the lizard, activating its body, and drawing it to the surface, only for frost to return? Reptiles have some ability to move in and out of hibernation but they can be caught out, their metabolism thrown into intense activity that cannot be reversed. In outdoor vivaria, lizards that emerge too early sometimes seem to get trapped between two states, too active to return to brumation, but not active enough or warm enough to feed. Before spring has come properly, insects are in any case hard to find. In these circumstances, lizards may weaken and perish. Bringing them indoors, into artificial warmth, sometimes revives them, but sometimes shocks their systems, seeming to demand that their bodies should change too quickly. They seize up, lose their power of movement and die.

What does the lizard know? Imagine knowing that these things might happen, and letting go, closing one's eyes, dozing off, leaving everything to hazard. Consciousness of time entails the ability to think about what might happen while we are sleeping. We assume that these creatures have no such consciousness. They go under each year without thinking about what may happen, or wondering whether they will see the light again. But I cannot help imagining things otherwise. These creatures, at these edge-moments, make me think of people at the moment of committing themselves to hazard, as we do, in a small way, when we sedate ourselves so that we will sleep on an aeroplane – or every time we feel ourselves drift off to sleep, anywhere.

My instinct is to want to be aware of my imminent ending – to be able to make my goodbyes, forgive myself and have some moments to look across my life; if only the fight-or-flight impulse can be suspended, so that I will not be entirely gripped by panic. If only. Most endings, I guess, are either oblivious or brutal: a panic of struggling for breath. Anything else would be astonishingly lucky. Of the two, who would choose brutal? It is a mercy not to know that these moments, now, are my last – not to see what is coming. Of course it is a mercy. But when I think of my father, I want to think of a moment of knowledge in which he sees he is loved and forgiven.

Airports, out on the runways, often have patches of wildness beyond the flat grass. Some have banks left to nature, and ditches of water, duckweed-covered. Strapped in my seat, as the plane taxies out, I am committed. I can't back out now. At this point, I look for those banks and ditches. They reassure me. I imagine frogs or snakes in them, animals to whom the plane is incomprehensible, a rush of sound and air too vast to notice – and yet they comprehend it all they need. Heathrow Airport once published a leaflet describing the wildlife within its perimeter. Grass Snakes were pictured, I remember. It makes sense, of course, because the runway areas require a large margin of space, and are so unapproachable by people. As my take-off begins, I imagine myself setting out, instead, for a day looking for those Grass Snakes, on those banks above dykes that recall the ones at Camber. With this thought I calm down, and imagine the bank at my feet as I wander along. It is a quiet morning. I feel safe here. The earth has dried out in the sun. There are cracks. A snake could emerge from one, and glide down to the water. If I sit here I might see it.

It was only a matter of time before snakes filled our thoughts. Snakes were the ultimate – the last frontier of reptile-keeping. They were the most charismatic reptiles, large, fast and menacing. It was always a dramatic moment when we saw one. The Adder was the most exciting of all, being actually dangerous.

Their movement is an important part of their charisma. They bring an expanse of grass or heather to life by racing through it in wide curves, too fast for the eye to see exactly where they are. The land itself seems to be moving. Venomous snakes withdraw their heads into the centre of tight coils and then lash out, hurling their gaping heads a yard or more. Pausing on the surface of a lake, in an S-shape, head raised, or draped at different levels in a tree, a snake has instant control over all of its wide-flung body. When it moves, the muscles at any part of the belly and sides can find purchase on tree or water. In seconds the snake whips away.

They give an absolute contrast of stillness and movement. Their movements are wonderfully pure, whether the snake is gone in the crack of a whip, or pulls its own thread carefully through a landscape, finding exactly where to go. They move like nothing else moves. The fastest snake that I tried to catch as a child was a southern

European species, black threaded with gold, with a fierce golden eye. It is now called the Western Whip Snake, *Coluber viridiflavus*, which is an apt enough name – the snake is shiny and black like a whip, seems to taper like a whip, often grows to five feet and can burst into lashing momentum. But I prefer the name this snake had when I was a child. It was The Dark Green or Angry Snake, and, indeed, on close inspection, the colour is an almost-black green. 'Angry Snake' was what I liked. Looking at the pictures, I could believe this snake was angry, a fiery whip of anger. 'Bites readily, hanging on with a chewing action' said one of the books, *The Young Specialist Looks At Reptiles* by Alfred Leutscher. 'Not very well disposed to captivity,' the book then said darkly, before adding, as a reluctant afterthought – one can sense here a long pause: 'Harmless.'

This looked like a fearsome snake, one we could barely imagine, and I did not expect to confront one, but it happened. I was with my family in Italy, on one of our camping holidays. Lake Trasimeno was the place. I slipped away into a maize field, between two tall rows. The field was like a forest. I went wading down a channel that ran through the maize stalks. The channel had bare earthy banks. On all sides the maize shut out the light, but the sun poured into that corridor. I was hoping for Grass Snakes, and indeed saw two elver-like small ones with pale yellow collars, looping off through the water before I could catch them. Then I saw it, on the bank to my right.

It was a pile of dark snake, coiled untidily. There was too much snake for the bank, and stray coils had slipped down the side. I saw the blue-black shining green, the yellow stippling. It crossed my mind that if I could get this snake home through the Customs – this possibly seven-foot spirit of anger – it would outshine every animal that any other boy could possibly be keeping. It would make me a reptile legend. And it was really only a foot from my hand. There was the head, I saw now, at rest on the grey earth, hard eye impassive. The snake did not move.

I could have caught it quite easily, I think. It was so close. But the size of this reptile unnerved me. Where would I put my hand? Perhaps the thought of that chewing action came into mind some-where. But it was more the sheer size. I didn't know how I could hold such a monster. Where I could put all those coils?

My hand went up and then stopped. Then I lunged, but my grab was half-hearted. To gather up this snake was so much against instinct. And then it was going, uncoiling itself in one movement, but the snake was so long that for seconds it seemed to be all around me, in all directions, disappearing into the water, but not gone yet, as my hands flailed this way and that. Then the patch of bank was empty.

So, yes, there is movement. But there is also improbable stillness. I recently asked one of the country's most eminent lizard-breeders why he did not have snakes in his collection. 'Well,' he said. 'The thing is, they don't do much, really.' And when they are tame it is true. They spend a lot of time sluggish, digesting.

Snakes can be utterly motionless for hours. In a museum in Lawrence, Kansas, recently, I saw two of America's most fearsome pit vipers. These are snakes with a 'pit' sensor, a declivity between eye and nostril that is intensely sensitive to heat and thus draws the viper towards its prey. This term is suggestive also of a familiar horror film standby – a pit of snakes. The Timber Rattlesnake was draped on a piece of bark: a thick grey and green snake, barred in black and turning completely black towards the tail. In the next vivarium, half submerged in its concrete pool, was the Cottonmouth or Water Moccasin, so called because of its warning display. It flashes the white inside of its mouth, and since the snake is a dark muddy green, it is sometimes the mouth alone that is visible on the dark water, looking like an open slipper. The Cottonmouth's head looks vindictive and purposeful, flat-topped with sharp sloping edges, as if made of metal.

Neither of these snakes moved once in ten minutes. I began to wonder if they were alive. There was a diorama of stuffed animals on the floor below. But reptiles, snakes especially, are rarely stuffed convincingly. Could they be models, perfect plastic replicas? Was it possible? *Please do not tap the glass*, said a sign above the row of glass tanks. Children must find it very hard to obey. I could see every fine detail of these two snakes, but without movement they barely seemed real. The minutes ticked on. I didn't want to leave, because any minute one of these snakes might move, but eventually, with a sense of disappointment – of something that hadn't quite happened – I had to walk away. There was nothing else to do. The charisma of snakes is in the dramatic moment of seeing them move in the wild. It is lost if you have them trapped in front of you, especially if they are motionless, as they are for long periods of time, digesting food or lying – as perhaps they imagine – in ambush. We were not experienced enough to realise this, though I think we began to perceive it, for what we imagined and talked about was always the moment of capture or the moment, with the snake in captivity, of creeping up and seeing it unobserved.

At first, we only thought of catching a Grass Snake. This thick green snake is the commonest British species, and the largest. Two and a half feet is a normal size for an adult. Three is unusual but not astonishing. The British record is five feet ten inches, and rumours abound that there are snakes of six or seven feet living in inaccessible marshes in forgotten folds of Britain. It is not impossible. This sort of size has been recorded in southern Europe, and the only reason Grass Snakes do not frequently grow that long is that the bigger and longer they are, the more vulnerable to predators. An animal whose main defence is to shoot into hiding as quickly as possible is at a great disadvantage having so much to pack away.

The name Grass Snake suggested to us a wide-ranging creature, one not restricted to a particular habitat like heath or wetland, but to be found wherever there was thick grass, throughout the meadows, downs, riverbanks and orchards of England; even perhaps in parks, gardens and cemeteries. It was a muscular snake, full of strength and gliding energy; able to skim across the ground in wide swerves with its head held high. They were, in fact, hard to find — but there was an exciting sense that in the countryside one might find one anywhere.

I remember holding one for the first time — snatched up as it raced from my feet on a walk with my family. We were walking through a field of cows, not a place where I expected to see Grass Snakes. Perhaps this was why I caught it. The snake took me by surprise and was in my hand before I could think. If I had seen it from a distance and tried to creep up, I might have hesitated too long. Holding the snake up high, I gave a whoop of triumph. My family were pleased for me, I think.

In the air the snake tried to find purchase as it dangled from my hand, pulling down with surprising strength. I brought it close to my face, and saw that the tongue was shiny black and grooved, as if two tongues had been stuck together and then pulled apart at the tip for the fork. In and out the tongue flickered.

The round eye was bold. Black markings were clean and precise. Tiger stripes went down from eye to mouth, on a yellow-white background. On top, the head was the green of green olives, the basic colour of the whole upper body. Behind the head was a collar of buttercup yellow — a crescent of yellow each side of the head, edged behind with a crescent of black. The old name in Britain was Ringed Snake, because of this collar. *Natrix natrix* is the scientific name.

Dull green with flashes of yellow – these are the colours of marshes, wet meadows and meadow-corner ponds, where reeds and water irises grow, and water buttercups. The Grass Snake is a spirit of these places.

It hunts frogs, toads, newts and fish there, also eating earthworms and occasionally baby mice and infant birds. When it has eaten, there is an elongated bulge that slowly travels down the body, pushed by the muscles of the stomach. Each vertebra has a rib protruding from each side, and curling down and around almost to meet the other, forming the tube of the body. The ribs are attached to the tube's lining of muscle, and because they are not connected to any other bones, the snake can open and jiggle a pair individually, to swallow and to move itself along.

Snakes have backward-curving teeth to grip their prey and inch it down into the throat, but they do not chew. Their jawbones are not joined at the point of the chin, and are loosely attached to the skull; each of the four jawbones can splay out from the skull and move independently. Because of this, the snake can engulf prey that looks impossibly large, forcing it bit by bit down the throat, where it is squeezed into a long shape for swallowing, but not necessarily killed for some time. The snake's skin stretches to accommodate the meal, revealing purplish skin normally covered by scales.

Once we caught a Grass Snake with a hefty bulge, and in our hands the snake began to choke, so it seemed, and whip its head from side to side. In a short number of convulsions, the bulge moved back up the Grass Snake's throat and into its mouth. A pair of feet appeared, frog's or toad's. The legs appeared, clearly a toad's. And then, as the snake shook its head, the whole toad slid out backwards, like a baby, immediately swelling to a size that looked impossible for the snake to have swallowed. The toad seemed quite unharmed, not even dazed. 'Well, it shows there is always hope,' said Adrian, finally.

My family gathered to look at my first Grass Snake.

'Ew,' said Cathy. 'It's peed on your arm. '

The snake had discharged a pungent liquid from a gland at the edge of the anus. It had emptied its anus as well. I wiped the chalky lump off my hand onto my jeans. This is one of the Grass Snake's defences. A fox or badger will sometimes drop the snake because of the bad taste or smell, giving the reptile a momentary chance to disappear. But we liked the smell, a musky, sweet and salty tang that lingered on our hands, and we would smell each other's held-out fingers in the fellowship of it, the smell and thrill of having handled a wild Grass Snake.

I didn't have a container for the Grass Snake with me, so I vowed that I would hold it, letting it twine around my arm, for the remainder of our walk.

'Are you really going to keep it?' asked Cathy. 'It's huge. What does it eat?'

'Frogs and newts, mainly, and fish.'

'You can't feed it on frogs.'

'Yes, we'll put frogs in for it.'

'Oh, Richard, you can't. The poor frogs. You're disgusting.'

And this *was* a difficulty, though I laughed it off. Soon we had five Grass Snakes, two at my house and three at Adrian's, and were planning to put them all together the next spring, in the hope that they would form a mating ball, which happens in the wild when a female's pheromones, released onto the air by her skin, attract several males at once, and they all twine around her. Her skin at this time develops a sheen of fine oil. It is not dangerous for the female, like the toad-ball for the female toad, but she is held in place while the males swarm upon her, engaging in tail-wrestling rather like human arm-wrestling, as each tries to hold his cloaca in place so as to insert a hemipenis. If you find one of these balls, the snakes are likely to be so preoccupied that they will not move away. Once they are startled, the ball unravels with oiled speed. Snakes disentangle and stream off in all directions.

Every week we went hunting for Common Frogs to put in as food. We would tramp through the long grass in the allotments until a frog jumped away from our feet, and lift every old barrow and wood-pile we could find, returning home with a cloth bag full of frogs that we would tip into the snake enclosure. Day by day, they would disappear, and bulges would appear in the snakes. I was rather aghast when I saw a snake feed. The frog was moving in short jumps across the floor of the enclosure. Suddenly the thick snake was there, as if thrown like a lasso. With a wide grinning gape, its head shot at the frog, which was caught by one back leg. The snake then pulled the frog to the shelter of a tussock, and began to swallow, working its head up the leg in biting swivelling grabs, each one taking more of the leg out of sight.

Grass Snakes do not hold their prey in their coils, but the snake's body gathered loosely around the frog, which was kicking its free foot ineffectually at the snake's head. Reaching the top of the thigh, the snake's mouth widened, shimmied a little, and gulped forward, forcing the other leg up alongside the frog's body, sticky with mucus. At this point the frog gave a long melancholy yawp, changing timbre into a groan. Its mouth was gaping. The sound was perhaps not a scream of distress – the books said that it was produced by air forced out of the body. We found it hard, though, not to hear feeling in the call, and we didn't like putting the frogs into the enclosure. It was a relief when we had the idea of buying goldfish from the pet shop. These were one-shilling goldfish that always disappeared from the water bowl in a day or two.

The year after catching that Grass Snake, I turned thirteen. Ade and Phil were fourteen. They talked a lot about the changes taking place in our bodies. Being one year younger, I was jealous but also relieved not to have to compete. At Ade and Phil's school, the showers after games were an opportunity for the boys to glimpse each other's development. This boy had a lot of pubic hair and swaggered around showing it; that boy had none yet and was embarrassed; another had none and was considered unlikely ever to grow any. These developments were discussed by Ade and Phil in tones of fear and wonder, and I listened with excitement. At one point, in his bedroom, after we had fed his tree frogs, newts and lizards, Ade produced from his bedside drawer a little ball of brown hair, tangled and curly, like the balls of hair that got caught in the vacuum cleaner. This was proof, he said. He had proper pubic hair, not just a few wisps, like a boy called Ellison who had been laughed at in the showers at school. That very morning he had cut off this tangle to show us. Tomorrow he would take it to school. The removal of this cluster had made no difference to the appearance of his thick growth, so he assured us. We passed it round like something for the nature table. Nervously I touched it with a finger.

We had started going to parties, and Phil had held one at his house. To start with, at these parties, boys and girls would gather at opposite ends of the room and talk intensely among themselves, casting furtive looks at the other group. After an hour of this, the two sexes would mingle a little, and some dancing would begin, and then couples began slipping out to the bedrooms or the rooms with sofas, for what we called 'groping'. I felt painfully young and embarrassed and usually got quietly drunk, sometimes taking my can of beer into the garden to sit looking at the stars and listening to the music, not unhappy, exactly, but glad to be out on my own. Ade and Phil would afterwards discuss how far they had got and what they thought of particular girls. One, Karen Carey, was very exciting but frightening. She wore red lipstick even in school uniform and held us appraisingly in her gaze; another, Jill Bligh, was posher and quieter. I fantasised a little that she might be bookish. These names come back instantly now.

At one of these parties, I was alone in the garden, and Ade and Phil came out to find me with a girl. Hazel Watts was her name. 'Kerridge,' they called out. 'Hey, Kerridge. Where are you?'

From the darkness of the lawn I went back up to the concrete yard, into the light from the kitchen.

'She's promised to kiss you,' they said. 'She's willing.'

As they stood round, she came up and placed her hands on my cheeks, drawing my face towards hers. Her hands moved to the back of my head. I felt her warmth. There was perfume. Her lips came onto mine and began rotating. There were no tongues. My nose was in her hair.

'Whoah, Ker-ridge,' they shouted. 'Ha-hey. His first snog.'

Phil said sorry to me later, but I was happy.

Another memory from that time is of a strange boy's front room. Trevor Perrot was his name. He went to Ade and Phil's school. Why we were there I don't know, but there were three girls too, and we were listening to records. A packet of chocolate digestives went round. It was a long wet Sunday afternoon. One of the girls was Susan Obee, Perrot's girlfriend, and she and Perrot were on the sofa. While we were talking they kissed now and then. We looked away, and our conversation drifted back to reptiles. The other two girls talked about something else, not looking at us much. I became aware that Perrot and Susan Obee were now grappling rather than kissing. He was undoing the buttons of her shirt, and she was laughing but trying to bat away his hands. 'Not here,' she said. 'Stop it.' We were all watching now. He had pushed her shirt back off her shoulders and half down her arms. We saw the white straps of her bra and the shiny, surprisingly stiff-looking fabric of the bra itself. Perrot pulled the bra up, squashing her breasts, and then, just for a second, the bra was off and we saw them, before she hunched herself over and ran from the room. 'I'll go after her,' one of the other girls said, flatly. The third one followed. There was a silence. 'I'll see if she's all right,' said Perrot finally and left us there. It was a long time before they all came back into the room together, as if nothing much had happened.

We didn't talk about what we had seen.

As we caught grasshoppers in the allotments, Ade and Phil discussed girls a lot, and sometimes sex. But, also, they disapproved of these subjects. 'Stop thinking about females,' one of them would say. 'We have to think about the animals. Are you coming to get insects tonight, or lying around thinking about girls?' Each of them did this to the other. They seemed to find relief in turning back to our collection, but then each wanted from the other a show of loyalty to the zoo. They wrote out a document in mock-medieval style with a drop of red wax at the bottom for a seal, a pledge that they called the Charter.

We the undersigned hereby vow that we will not be distracted by females, but will be true to the noble reptile hunt, and not rest until we have all the British species of reptiles and amphibians. If one of us is in danger of breaking this pledge, the others will produce the Charter and at the sight of the Charter the tempted one will come to his senses immediately.

The three of us signed, but I think my signature was only wanted in order to disguise the fact that this was something they were requiring of each other.

Phil had gone on holiday with his family to Christchurch, on the coast near the New Forest, and one day had wandered off from the beach looking for reptiles. On the wasteland of brambles and bracken behind the beach, there were lots of Common Lizards and Slow Worms, still exciting to see but not needed for our collection. We already had fine specimens. Something made him go deeper in, he said. He fought his way through the bracken. At the back of the strip was a fence made of thick boards. Behind it were fir trees; in front of it, the bracken had been cut down. In the stubble lay a large board, like the ones in the fence. A lizard ran off the top and went under. Phil squatted in front of the board, and carefully raised it with one hand. Beneath, neatly coiled in a long thin loop, was a small brown snake. Phil did not realise what it was.

He had found the Smooth Snake, *Coronella austriaca*, the very rarest of the British reptiles, harder to find than the Sand Lizard, whose southern range it shares almost exactly, though it survives in the New Forest where the lizard has died out. Not only is the Smooth Snake extremely rare in Britain, but the ones we do have are almost always out of sight. Smooth Snakes do occasionally emerge and glide through the heather to hunt, but most of their time is spent in the heather roots or anywhere under cover that is warm. The way to be sure of seeing them is to leave small sheets of corrugated iron on the heath where these snakes are known to live. In the sun, the metal heats up, and the snakes come and rest under it to gain the body temperature they need to incubate their young, and digest their prey.

Smooth Snakes have been known to live more than thirty years, and feed almost exclusively on other reptiles, including the rare Sand Lizard and occasionally Adders. This is the only British constrictor – a snake that subdues its prey in its coils by squeezing. Rees Cox, for many years warden of Studland Beach Nature Reserve, tells a story of how a research team inserted chips into Smooth Snakes in the reserve, tiny radio transmitters that would reveal the movements of these reptiles that rarely bask and live mostly out of sight. What was puzzling, immediately, was that two snakes seemed inseparable. They went everywhere together. The two signals always came from the same place. But of course what it meant was that one snake had eaten the other.

Another story about radio tracking is that researchers studying Sand Lizards on a patch of heath in the suburbs of Poole in Dorset found that the signal from one lizard led them off the heath and down a residential street, and then another. The signal got stronger as they approached a particular house, and faded as they moved away. A woman answered the door and admitted that the lizard was indeed inside the house, in a fish tank. Her young son had brought it back from the heath.

Smooth Snakes eat a great many Slow Worms, which are easier to catch than legged lizards and other snakes, but this does not deter the Slow Worms from joining them under the corrugated iron, sometimes in large numbers. The warmth under the metal is irresistible to both species. Smooth Snakes are a pale, slightly washed-out or greyish brown in basic colour, marked along their bodies by lines of darker brown flecks. They have a darker brown patch on the top of the head, and a horizontal dark line that seems to run through the eye. Sometimes the patch on the head is in the rough shape of a V, which can lead to confusion with the Adder and thus to danger for the Smooth Snake. The small eye is yellow-gold, with a round black pupil. It is a beautiful eye that catches the light and shimmers, and although the snake's basic appearance is plain — it is the faded colour of dried out, bleached-out, soaked-out bracken leaves — it often has a delicate gingerish tinge, especially in the males, and the grey underside too is sometimes interlaced with red or ginger. As the name suggests, this snake has a smooth body that in the sun, or when the snake has just sloughed, can look burnished. Sometimes there is a tortoiseshell gloss. The flickering tongue is pale red.

We installed the Smooth Snake in a long fish tank in Ade's bedroom, warmed by two forty-watt bulbs. Ade arranged logs and bark on a leafmould base. Now, having solved the problem of feeding frogs to the Grass Snakes, we had to catch lizards and Slow Worms for the Smooth Snake – something we found just as disagreeable, if not more so, though they didn't have the naked moist helplessness of the frogs.

Phil brought this snake back for our collection, and it deserved pride of place. It was certainly the rarest animal we had. But being excited by the Smooth Snake took specialist knowledge; it was a snake connoisseur's snake. We had now set our sights on an animal much more charismatic. There was one reptile that would impress everyone who heard about it – a reptile that would arouse the thrill of danger and make us seem cool and dangerous as well. We wanted an Adder.

At that time our assumption was that the bite could kill you, and we were not *quite* wrong, though death from Adder bites is extremely rare. There have been none recorded in Britain for thirty years, though about ninety people a year are bitten, usually when they inadvertently touch the snake or attempt to handle it, like the couple walking in Scotland recently, who found an Adder and decided to photograph each other holding it, Steve Irwin style. Apparently, when his photoshoot was over, the man simply handed the snake to the woman, who was bitten. Why hadn't the council warned them, they wanted to know. A young child might be in danger, or someone with heart trouble or an allergic reaction, but a bee-sting is much more dangerous.

A girlfriend of mine was once bitten spectacularly. We were walking across heather. I had brought her to Studland with the intention of showing her some reptiles, and of showing off my own poetic love of them, wanting really, as well, to find out if she could see their brilliance at all. It was April, and the Sand Lizards would be at their greenest. I led her down a narrow ravine that was over-hung with heather. There was a place at the end, a patch on a bank, where I had seen a nice male the week before. We never got there.

'There's something here. Oh God,' she shouted. 'Fuck!'

She was hopping, her wellington boot in the air. I saw that from it was hanging a small light brown snake with a zigzag line. It was twisting itself madly, and I realised it was stuck to her boot. The mouth was biting at her boot. I saw a flash of the mouth's pink tissue. Presumably my friend had trodden on the Adder, on the path beneath the heather, and the snake had struck its fangs into the rubber of her boot. Now it could not disengage. One fang, at least, was snagged. As her foot waved in the air, the snake's body coiled around it.

'Don't put your foot down,' I said.

'What?'

'Keep your foot in the air. If you put it down, you'll crush the Adder.'

'It's an Adder? Shit! Get it off me.'

'OK. It can't bite through your boot. Lean on me, and I'll support your leg.'

'Fuck you, Richard.'

But she did lean on me and I caught her thigh in my arms. We watched the snake knot itself tighter and tighter.

'I want it off me.'

'OK. Give me a minute.'

The Adder was now a tight ball on her black rubber boot. I was holding her thigh and her calf. We staggered a few paces. Abruptly, the snake dropped off.

We both looked. Carefully, I set her foot down on the ground and stepped back, unsure what to do with my hands. We now stood apart from each other.

'Can you forgive me? I never meant for this to happen.'

'I don't know,' she said.

I was bitten myself once, I *think*. Deep in daydream, I was walking across the same heath wearing trainers, assuming that snakes would flee from my approach, since it was so difficult to see them when I was trying to do so. Plunging my foot down into the heather, having forgotten the possibility of Adder-bite, I felt a sensation like a hot needle going slowly into one of my toes. *Adder* didn't occur to me even then: 'Ow, what a nasty thorn,' I thought, having seen no snake movement in the heather. For the rest of the day I felt discomfort, except that when I touched the toe on something, it felt very sore.

221

So, that night, I was unprepared for what I saw when I removed my sock. The toe had turned almost entirely black, a characteristic symptom of Adder bite. But there was no swelling or continuing soreness, and the blackness quickly faded. It was probably an Adder, but I cannot be sure.

The traditional remedy is to catch an Adder – it doesn't have to be the same one – slit it down the belly, fry it, and cool the fat that comes out, which is then used as an ointment. Alternatively, Thomas Pennant reported in 1772 that the natives of Isla believed a poultice of human ordure to be an infallible cure. There were also some benefits that might be had from Adders. Consuming them was thought to cure not only the effects of the Adder's own venom, but other cruel afflictions. This passage is from *De Proprietatibus Rerum*, by Bartholomew Angelicus, written circa 1260:

> . . . to heal or hide leprosy, best is a red adder with a white womb, if the venom be away, and the tail and the head smitten off and the body sod with leeks, if it be oft taken and eaten. And this medicine helpeth in many evils; as appeareth by the blind man, to whom his wife gave garlick with an adder instead of an eel, that it might slay him, and he ate it, and after that by much sweat, he recovered his sight again.

Here, again, is the ambiguity of the serpent.

In Britain it is rare for serum treatment to be needed, but anyone who even suspects they have been bitten should get to a doctor. Fatalities may be rare, but the swelling and bruising can be extreme. The demonisation of the Adder in traditional British culture is more understandable than that of the toad, but still unnecessary – and attitudes have changed to the extent that serious thought is now being given to methods of conserving and increasing Adder numbers, which have declined severely. Adders are still found in most parts of the country, on heath and scrubland, woodland verges, rocky quarries, downs and moors and mountains. In all of Scotland but the most southern parts, it is the only snake you will see. But colonies are more isolated, and in many places the habitat corridors enabling snakes to make contact with other populations have disappeared. Tests have shown that many colonies are suffering from loss of genetic diversity. Defects are appearing that threaten future breeding, and Natural England is considering a trial policy of swapping snakes between colonies, to mix up the gene pool. If this practice is intro-duced, the Adder in Britain, like the Sand Lizard, will become an artificially maintained species, there because we want it, a work of art – in this case, art in the Gothic genre.

I have sat a few feet from a pile of bramble-covered rocks and seen seven male Adders threading frantically through them, driven wild by the pheromones released by the oily skin of a female. She was large and ginger-brown. They were smaller – some of them very small – and light grey with shiny black markings; the largest was black all along his lower body and under his jaw. He was following the female closely, his head at her flank, his tongue flicking onto her glistening scales. She pressed on ahead of him. Other males followed, and one approached the female from a different angle – a pale snake, almost white with black zigzag. The dark male whipped his head across her body, and for a moment the two rivals sparred, twining around each other while still moving fast behind the female, each trying to force the other to the ground and hold him down. They did not rear up while doing this, into the famous Adder 'dance', but it was nevertheless a dramatic moment, before the white male broke off and turned away. The dark male then followed the female, who slowed and waited for him. On a mossy bed, safe under a bramble, their bodies became entwined, the male touching her all the time with his flickering tongue. Their intertwined tails began jerking, increasing in speed, reminding me of the buzzing tail of a rattlesnake. The male had found her cloaca with his hemipenis. She moved now, uncoiling, spooling down the bank, through the brambles and across the stones, and he was dragged behind her, attached firmly, his body awkwardly across hers and snagging.

How would we catch one? We had considered this question. An Adder picked up by the tip of the tail would not be able to bring its head up to bite, Phil announced. You could safely dangle it into a collecting bag. This he had discovered in an old book, written, it said, for boys who loved adventure. But don't pick up baby Adders, the book warned, for these *were* capable of curling up to bite you and were full of poison. For adult vipers, the method was safe and better than grabbing behind the head, because a snake held at the neck might edge its head sideways and put a fang into your finger. We nodded wisely.

Mind you, some of those books seemed strange even to us. *The Junior Naturalist's Handbook*, for example, by Geoffrey Grainge Watson, a book aimed at young boys keen on nature, contained step by step instructions for skinning a lizard and preserving the skin between two panes of glass. Did many boys actually do this, I wondered. After seeing the boy with the squirrel in the library, I wasn't so sure that they didn't. This was not one of the older books, either. 1962 was the year of publication – just six years before.

When we next met, Adrian produced an old biker's glove, a thick leather gauntlet that surely the fangs would not pierce. But we only had the one.

In our largest zinc bath, we planted tussocks and arranged logs, working at creating the look of a wild place where adders would bask in the hot stillness; a place where we would creep up through bracken and see them, and catch our breath. Again, we stretched net curtain and wire netting in a wooden frame, with foam rubber glued to the underside. We felt we were experts at this now. The Adder would be our zoo's tiger. It mustn't escape.

Then, on a hot May morning we set out, cycling through the suburbs and freewheeling headlong down the great chalk drop into the Weald Valley, aiming for the heaths of Ashdown Forest. It was 1968. I was thirteen.

And then, just yards from the road, everything went still. I was poised above an Adder, which looked up at me, its tail-tip wriggling like a worm. The furious little face was oily black. White scales like tiny pearls lined the top of the mouth. Adders have faces intense with hatred, hot with it. The eyes were like blood-blisters.

What had caught my eye in the heather was the zigzag, a pattern too clear to look natural. The shadows cast by bracken leaves have similar shapes. In these shadows, the zigzag evolved, presumably, but somehow the scatter of light and shade on the forest floor became on the snake a regular wavy line. It breaks up the animal's outline. Hawks and crows see the snake from above. People do too. When the snake moves, winding through stalks and shadows, the zigzag goes in different directions, confusing the eye. On a motionless snake, it is insolently clear. In the heath's debris, the zigzag looks stylised, like a printed or ceramic pattern, a logo or uniform, a badge of power and purpose. When we thought of Adders, the other creatures in our zoo seemed weak and flustered – the little biscuit-coloured lizards and soft, gaping frogs. They were low status. An Adder was deadly cool.

The zigzag is common to all Europe's vipers. It is one of nature's design classics. On some, such as the Nose-Horned Viper of south-eastern Europe, the colours seem freshly painted. Chocolate on pale grey, scarlet on pink – a touch might make them smear. The

Nose-Horn comes from a land of white light and black shadows. In comparison, our Adder, the Adder or Northern Viper, has muted colours. Females are the assorted browns of dry grass, old bracken, leafmould and sand. Males are light grey marked in black. Rarely, the males are white with black markings, and the females ghostly pink. This viper of Northern Europe lives under changing skies. Shadows cross the heath like changes of mood. Browns and purples intensify, then lighten. The Adder's colours sink into the background. Then, at the sun's touch, they gleam.

'Here's one.' I kept my voice low. Adrian threw the glove, which fell short in deep heather. I didn't dare move.

'Grab the tail.'

'I'll try.'

I dropped to my knees, and forward onto my hands. The moss was spongy. A sour smell came up. My thumb felt a thorn. Inching until my head was almost over the Adder, I raised my right hand.

Something touched my left.

I looked down and gasped at the huge ginger Adder uncoiling beside my hand. Her pale yellow head had an imp's face with copper eyes. The touch I had felt was her nose on my hand. She approached again, black tongue flickering.

Her eyes burned. The pupils were black and vertical. In a male Adder's dark red eye, the flame is a flare in deep space, far off. His fury has distant origins. The female's lighter eye burns closer, with intimate malice.

She paused. The tongue was now fully extended, trembling, its tips thin as hairs. At the root it was pink.

A slow hiss started, as her body filled and emptied.

There was something about the paleness of this head. For a moment, it looked like a whole creature, an insect, something wasp-like, or a fat insect revealed by the lifting of a stone.

People used to talk of the Adder's 'sting'. It is in Shakespeare and the King James Bible. Now we use 'sting' for the lance on the tail of a wasp, or the thorn a bee leaves in the flesh. An Adder bites; it has no sting. But the word 'sting' has some visceral effect. 'Bite' is from the biter's viewpoint, but 'sting' suggests the feel of being bitten – the spasm and the pain spreading out. I looked at the Adder. Her bite would be like a dentist's needle.

But her head turned away. She was gliding into the heather. My fingers grasped the disappearing tail. It felt warm. I pulled her into the open, and stood up shakily, as she twisted and bucked, hissing loudly. She was heavy. Her tail tip coiled round my finger. She threw her head at me, mouth open; I staggered and swung her away. Her underside was reddish grey. She whipcracked, wrenching her body, then the head came towards me again. I stepped back, nearly falling over. 'God!' gasped Adrian, coming up behind.

I didn't know what to do.

The face of a snake is unmoving. No affection, doubt, surprise, curiosity or fear comes into that countenance. It knows everything it will ever know. The eye, behind a hard, transparent scale, does not move or blink.

Adders look wickedly intent. Big flat eyebrow-scales frown towards the nose. The mouth has a long sad line, or an evil grin, or a glum smile as if Adders hate being Adders and want revenge on everything. They would cry if they could, but their eyes are too hard, in their medieval demon faces. Devils are unhappy, shut out of paradise, but do not know how to be other than devils. This yellow face knew something that everyone knows.

Most snakes are ambush predators. They *lurk*, motionless and camouflaged, waiting for unsuspecting prey. Then they carefully choose their moment to strike. Or perhaps I should say that their eyes, heat-sensors and receptors of airborne chemical traces trigger the strike. But it looks like choice, and the human behaviour it seems to resemble is calculating, spying and manipulative. Add to this the connotations of killing by injection. 'Cold-blooded' is a term describing a physiological survival process, but we also use it to express the idea of 'cold' rationality, as opposed to the 'warmth' of human sympathy. Snakes give a glimpse of a world with no pity. Stone faces will ignore our pleas. Our vulnerabilities will be exploited, not forgiven.

Yet these features are nothing more than survival adaptations or evolutionary quirks. The hard scale covering the eye is thought to be a protective screen, which evolved either in aquatic reptiles that later came out onto land, or lizards that took to living underground and lost their legs before becoming surface creatures again. These are the two main theories about the ancestry of snakes. There are evolutionary explanations, likewise, for the vertical elliptical pupil and long sad mouth. As a partly nocturnal predator, the viper benefits from the sharper image this pupil casts onto the retina, and the wide gape has evolved for the swallowing of prey, since the teeth are too weak to tear off flesh. In fairness, we should see the Adder as a sensitive creature, exquisitely alive to its environment in ways our senses cannot register; a vulnerable creature, frequently broken, and no more or less devoid of sympathy than most natural predators. We should. But this is like objecting to vampire films because the vampire is an unfair representation of a human being, an example, literally, of demonisation. It is a serious moral objection – and it fails to consider the appeal of such demons, the needs that they meet. The question should not be why the snake looks so evil, but why we have associated evil with these particular features. Why do we see the snake's face as a devil's face? And why do we need devils?

The biologist Edward O. Wilson, the great founder of 'sociobiology', or 'evolutionary psychology', as it is more often called in Britain, is also famous for his theory of 'biophilia'. This is the thesis that human beings have an instinctive and essential feeling of natural kinship with non-human animals. Cut off from those animals, we are unhappy, deprived of a vital part of ourselves. Snakes are one of Wilson's great examples. He is a snake-lover himself, full of snake stories from his Florida childhood. But snakes are his most paradoxical example, since they more obviously provoke phobia than love. A horror of snakes seems to be one of humankind's great evolutionary inheritances. Another evolutionary psychologist, Roger Ulrich, describes experiments in which people were repeatedly shown images of snakes and spiders, and also of dangers belonging to present-day industrial society, such as handguns and frayed wires. As the images recurred, the fears provoked by the modern items faded quickly, but the snakes aroused horror time after time. There was no getting used to them. Even in modern Europe, where the actual danger from snakes is vanishingly small (bees kill many more people, and cars are right off the scale), the terror is still common.

According to Wilson, psychological studies have found snakes appearing in human dreams more frequently than any other animal. In many cultures they are worshipped. Ulrich suggests that the shrines and rituals may be a means of coping with the anxiety these creatures arouse. Many traditions have snake-gods, some evil and destructive, some wise and benign, and some ambivalent in their attitude towards human beings – they are dangerous but they can become allies. Snakes have been used as symbols for many conflicting forces: death, life, creation, rejuvenation, fertility, eternity, infinity, temptation, treachery, sexual arousal, the phallus, the female sexual libertine, wisdom, knowledge, order, healing and medicine.

In Christian culture, the snake is present at the beginning of the human story, as the catalyst that brings about the great defining crisis between God and the first human beings. The serpent in the Book of Genesis persuades Eve to eat the forbidden fruit of the tree of the knowledge of good and evil. 'Your eyes shall be opened,' says the serpent, 'and ye shall be as gods, knowing good and evil.' The consequence is the Fall. Human beings are cast out of paradise into a fallen world, becoming mortal. Their knowledge of good and evil is their power, their pride and their burden. They have become beings of conscience, who agonise over questions of right and wrong and consider themselves responsible for their own behaviour. This, they believe, is what distinguishes them from animals. In their new self-consciousness, they become prey to thoughts of how they may look to others. Shame is born. 'And the eyes of them both were opened, and they knew that they were naked; and they sewed fig leaves together, and made themselves aprons.'

Thus the snake is midwife to the birth of the human condition. It is also the evil genius who plotted the Fall, since the serpent in Genesis is traditionally Satan in disguise. Adam, Eve and the serpent receive their punishments side by side, and are cast out of Eden together. To the serpent, God says, 'Because thou hast done this, thou art cursed above all cattle, and above every beast of the field; upon thy belly thou shalt go, and dust shalt thou eat all the days of thy life: and I will put enmity between thee and the woman, and between her seed and thy seed; it shall bruise thy head and thou shalt bruise his heel.' The snake is to be the eternal enemy of human-kind, but it has been our co-conspirator, in the dock with us. When we look at a snake under the influence of Christian tradition, we look at our enemy but also our old confederate. The snake reminds us of our guilt, but also our rebellion and defiance.

For the anthropologist and evolutionary psychologist Lynne Isbell, the snake's role in this story is not surprising. In her view, the Genesis story is the inscription in culture of deep evolutionary history. Knowledge of our long-term relationship with snakes comes to the surface in that story. Isbell's theory is that the presence of snakes as the main predator was the most important evolutionary pressure that developed the visual organs of the early primate ancestors of human beings. This is truly deep evolutionary history, going back further than the emergence of *Homo sapiens* and the long period of hunter-gatherer life on the African savannah that is the basis of most evolutionary psychology. Isbell points out that snakes were the main predators for the first small placental mammals and the anthropoid primates that followed them. These were stages in the ancestry of hominids. Birds of prey and carnivorous mammals evolved much later, when the physical structures human beings later inherited were already essentially in place. Snakes were the first great antagonist.

Because of snakes, Isbell says, evolutionary selection in our ancestors favoured clarity of perception in the lower visual field. You had to be able to spot the waiting snake before you reached it.

Eyes moved to the front of the head. The snake's eye is hard and chilling, but it seems that our own soulful eyes and full direct gaze have come about through our prehistoric relationship with this animal. Eyes that meet ours in recognition are a definitive sign for us of human sympathy and contact. Isbell says that those eyes were formed and positioned, and our faces arranged around them, because of our need to see snakes, whose eyes have the opposite meaning for us. The snake is everything inhuman; the embodiment of a cruel nature from which our humanity strives to redeem us. Yet, because of our evolutionary relationship with this creature, it is deeply familiar to us, awakening in us some of our most primitive and formative feelings, and thus bringing us back to ourselves.

Glimpsing a snake in the undergrowth ahead, our distant ancestor needed to recognise what it was and respond instantly. Neural pathways developed between those frontal eyes and the brain, so that the messages would reach the brain faster and trigger bodily reactions, such as floods of adrenaline, more quickly. In response to the presence of snakes – who were themselves being selected for effective camouflage and stronger venom – the visual organs improved until vision became the main sense. This led to a change of diet, a turn from insects to fruit as the staple food. Brightly coloured, strong-smelling fruit could be found without the keen sense of smell necessary for insectivores. Hence – according to Isbell – the weak sense of smell in hominids, and the reliance on vision.

With these developments went the improvement of visually

guided use of the hands, and the emergence of pointing with the finger as a means of communication. Isbell suggests that snakes were the main danger that needed urgent pointing-out, and therefore the main evolutionary pressure that led to this development, too. The need to turn rapidly in the direction indicated by a pointing finger prompted further changes to the structure of the brain. Evolutionary psychologists have suggested that this sort of gesturing played an important part in the development of communal behaviour, and was one of the main foundations of language. One theory, advanced by the Vietnamese linguist and cultural historian Huynh Sanh Thong, is that language grew especially out of the verbal and gestural signs with which mothers warned their children of the presence of snakes.

In other words, what human beings *are* is, to a surprising extent, the product of a relationship with snakes over millions of years. Isbell regards this relationship as the most important single factor in anthropoid evolution. For our small mammal ancestors, the factor was predation. Later, when anthropoid primates and early hominids had become too large to be the normal prey of snakes, it was the danger from the defensive behaviour of venomous snakes – still a major cause of death in many parts of the world. The only controversial part of Isbell's thesis is her attribution of such overwhelming importance to the evolutionary pressure from snakes. No one in evolutionary psychology denies that the pressure was at least a significant factor.

So it is reasonable to assume that intense responses to snakes are deeply encoded within us. Even if Isbell is only partly right, our organs of vision, the structure of parts of our brains, our habits of reaction, our sense of danger and our communication by bodily gesture and language are all full of the memory of snakes. Snakes are deep in our nerves, our brains and our genes. At the sight of a snake, an elemental part of us springs into action. When my eyes caught on that zigzag line in the heather, they and my mind were reacting to a stimulus that once shaped them. The release of adrenaline that made my heart surge was a flood of bodily memory, jolting through me even as I formed the thought *snake*.

We are full of the memory of snakes; primed to watch for them.

And here I was with a big one in my fingers, thrashing wildly.

Adrian held out the coffee jar. I swung the snake towards him. He leaped back, dropping the jar.

'Go on. It can't bite while it's dangling.'

'You reckon?' He put on the glove, and again raised the jar. I tried to manoeuvre the snake. Her snout touched the rim. Feeling support, she began to wind her body round the jar. Ade dropped it again. She corkscrewed. I felt her body stretch. Would the muscles tear?

Her head leaped at me again. I swung her away.

And her tail went through my fingertips. She flew, in an arc, towards Adrian, who jumped back. I saw her draped on the heather. Her head moved to point downwards, and she was gone.

Striking is a reflex. Certain signals prime the snake's body to strike, and then trigger the strike itself. They are various. Snakes are highly alert to movements above them, and, depending on how near and how high the movement is – swooping birds are a constant danger – they will whisk away into the grass or contract into a coil, the launch-pad for a strike. Body secretions in the air have the same effect. Snakes are very sensitive to these. Caught on the tongue and tasted in the mouth, these chemical traces tell the snake that prey or danger is near. Imagine the tingling of your own tongue much stronger. If the taste is prey, the snake will investigate; if the prey is approaching, the snake readies itself in ambush. An alarming taste makes it flee or coil defensively. Ground vibrations are messages too; a snake has no external ear, but vibrations felt by the underside of the jaw are communicated to the inner ear, and vibrations also register on the skin. The snake's organs are always making that calculation: flight or strike, and the snake just does.

If it can, a snake will flee from any creature large enough to be a threat. Usually, the various messages warn the snake before the creature is near; that is why it is so hard to see snakes. But sometimes they are taken by surprise. Colder temperatures make them sluggish. Females incubating young may be reluctant to move. Foraging Adders, or sexually excited Adders, may be oblivious to danger. If the snake is touched, especially if part of its body is pinned, and it has enough whip from a coiled position, it will strike.

The strike is a lashing action. The moment it is launched, the mouth opens wide, forcing the hinged fangs to spring upright out of the gums that wrap them, and squeezing the venom glands behind the eyes. Poison is forced along the venom ducts into the fangs and out into the bitten body.

Striking is what Adders do. The stimulus makes it happen, drawing the strike and the venom up out of the snake. There is no hesitation. This is what we mean when we say that animals have no conscious will. The snake doesn't decide to strike. Conditions occur that make it strike. I imagine the snake's mouth excited, salivating, itching to an unbearable intensity, until everything hurtles forward.

How people manage to strike quickly is a mystery to me.

Once I hit a squirrel with my car. Ahead of me, on the far side of the road, it ran out and stopped, a foot raised. A car, coming towards me, was bearing down on the squirrel. It tensed, flattened itself, made to turn back, thought 'there's no time', turned again and dashed under my wheels. A small thud, a wobble as the wheel passed over something; then I looked in the mirror and saw a grey shape twisting on the road.

I ran back along the verge.

It had dragged itself to the gutter, and lay panting, black eyes wide. I scooped it up onto the grass, its body warm. Hissing at me, it tried to lunge and get away, but the lower body wouldn't move. Head, shoulders, arms jerked helplessly, then stopped, and the animal looked at me quietly.

I've got to kill it, I thought, I must kill it, not leave it dying slowly. Looking round, I saw a tumbled-down wall, spilling bricks. There was a lump I could just lift – a slab of several bricks, still mortared together. That would do. Hefting the slab over, I raised it and brought it down to pound the squirrel's head, but as my weapon descended the animal somehow twisted and threw itself out of the way. Pure reflex, and only just in time – the slab almost touched it. Or had I held back? The squirrel was now on its back, next to the slab, face above white belly, like a child looking up from its bed. How had it found the strength? Could it do so again?

It didn't have to. I couldn't bring myself to try again. Pure instinct, no doubt, made the dying, paralysed animal find the strength to wrench itself out of the path of the bludgeon. But to me it seemed that the animal knew its predicament and was choosing. It wanted its final minutes of consciousness, even in pain by the roadside. And I couldn't bring the bricks down on that face. I left the squirrel there.

An hour later, most likely, it was stiff.

There are lots of them, I thought, as I walked away. But that was that one.

I felt cowardly. I had let pass the moment when it was possible to do what I had resolved to do. A familiar indecisiveness had stopped me.

Some of my friends did boxing at school. I found it appalling but saw how it taught them to hit without hesitating. They learned not to be distracted by the complicated signs, the eagerness and anxiety, passing across a looming face. You had to ignore those, and hit the face hard. No pawing – no half-hearted blows that were more like caresses. When the right moment came, you hit with all your force. That's what you had to do. If you saw those complicated feelings at all, they registered in some detached part of your brain, unconnected to your fists. This, I could see, was therapy for the doubtful. It taught you to commit yourself to the action, hazard yourself, make reality, not wait for it. But I couldn't. When I tried to fight, my punches were tentative. I didn't put all my force into them, and didn't know how much force I had. By the time my blows made contact I had slowed them, or changed their direction half-way. 'Perhaps we can just go through the motions,' was what those blows said. I don't think my opponents hit me properly, either. They didn't regard the fights as real.

One fight may not have been real as a physical contest, but seemed momentous between me and my father. I had never told him about the cycle ride to Camber, or the accident a few months later, when I was knocked off my bike and the car wheel missed my head by an inch. He continued to insist that I should ride only on the pavement. This was illegal, but he seemed unworried by that. 'It doesn't matter as long as you are safe.' 'All right, all right', I promised. 'I'll stick to the pavement.' Of course, I had no intention of keeping this vow. The embarrassment in front of my friends could not be contemplated. They were already beginning to tease me. I rode in the streets with them everywhere, rising up off the saddle as they did to pump my way up hills, and joining in the no-handed swoop on the other side, with arms stretched out straight either side.

Inevitably, my father saw me. His car came up behind us, the grey Rover, at the worst possible time; we had just finished a down-the-hill swoop. Perhaps he won't see that it's me, I thought; already I had the numb feeling that came when I was in trouble, the feeling of every muscle tightening. He passed us and stopped further on, and got out and flagged us down.

'Get off the bike,' he said.

'Look, Dad. I'm sorry.'

'Get off the bike.'

He was red in the face.

'I'll go on the pavement.'

He pushed my chest and grabbed the bike, pulling it from under me; then he raised it above his head, and threw it on the verge, front wheel spinning.

'Get home,' he said. 'Now.'

My friends thought I'd better obey him. They hadn't seen anything like this.

When I got home, he wasn't there. Mum said he wouldn't be back until evening. 'Is something wrong?' she said.

'You'll find out.'

'Tell me.'

'He saw me riding my bike on the road.'

'Oh, Richard.'

'But my friends all do. No one stops them.' I was crying.

'I'll try to talk to him.'

But he came in shouting, 'Where's that bloody boy?' I came down to face him. 'Has he told you?' Dad asked my mother. 'Has he told you what happened?'

'He said he was riding on the road.'

'Yes! Did he say he was riding no-handed? Showing off to his stupid friends! He'll get himself killed. You lied to us! You lied!'

'All right', I said. 'I'm sorry that I lied.'

'How could you do this, Richard? Can't you at least think of your mother?'

'It's hard for him,' said Mum. 'He's embarrassed in front of his friends.'

At this Dad exploded. 'His friends. That's the whole trouble. Those bloody friends. Next thing you know, they'll be getting motorbikes and racing at 90 miles an hour. Then what are you going to say – give him one of those?'

Dad was right about this, as it happened. Adrian and Phil were indeed beginning to talk about motorbikes a lot. There was one in Phil's street, a turquoise Bonneville, often parked in a garage forecourt. They stood and stared at it, making excited comments. It worried me a little. I was too young. My dad would never let me have one. I was scared of them, anyway.

'It's those bloody yobs that have influenced him and made him lie to us.'

'Shut up,' I shouted. 'Shut up!'

'Don't tell me to shut up!'

He swung his flat hand at my face.

And I punched him in the chest.

It was a token, a feeble jab, more like a prod. But he looked at me with horror. 'He's hit his father! That's it. He's hit his father!' He made as if to slam his hard fist in my face.

'Paul!' my mum shouted. But the fist stopped in front of my nose.

'He's hit his father. I can't stay here. What am I supposed to do? I can't live here with him. He's hit his father. I'm leaving. I can't stay in this house.'

The door slammed, and we heard his car start up and pull away. My mother was silent. 'You'd better go to your room,' she said at length.

'Will Dad come back?'

'I'm not sure,' she said quietly. 'I expect so. You both could have handled that better.'

'I don't want him to leave.'

She looked at me. 'I'm glad you say that, Richard. I'll tell him.'

Reading in bed, I heard his car come into the drive. I switched my light off. The front door opened and closed. There were footsteps on the stairs, then hushed voices that went on all night.

We did get our Adder that day. Phil caught it easily. We were walking beside a high bank. He turned his head, made a quick movement, and a snake was hanging from his fingers, as if taken from a shelf. 'Got one,' he said, matter-of-fact. Then his smile grew bigger and bigger. We gathered to look at a small female, russet-brown, squashed against the side of the jar. On her sides the scales were diamond-shaped, emerging from brown skin.

The bath was in Adrian's garden. Phil had already said he couldn't keep an Adder. His little brother might fiddle with the enclosure, and there were dogs that ran tumbling down the garden. Adrian's sisters were older than Phil's brother, and always kept away from his reptiles. He had no dogs. But we began to worry. Next door there was a little girl, a toddler. We imagined the Adder getting through the fence.

Our unease grew. After a week, Adrian told us he had begun checking on the snake all the time. One day he phoned, panicking; he couldn't find it. We took the terrarium to pieces. Out came every log, every tussock. The Adder wasn't there. Three lizards we had put in for food ran frantically about as we searched. They had seemed unworried by their room-mate, and the Adder had not eaten since its capture. Feeling scared, we searched Adrian's garden, finding nothing. 'What are you looking for?' called his mum.

'Oh, nothing.'

I went back to the bath. It had to be there. Lifting the lid, I noticed a narrow space between the planks. Wedged there was the Adder, tight and drab.

'We'll have to take it back,' said Adrian. 'Unless one of you can keep it. I can't stop thinking of it getting out and biting someone.'

The snake had grown in power. It had forced us to release it. We were quiet as we cycled to the heath. Shaken from the jar, the Adder rolled out in a ball. For perhaps a minute, there was no movement. The ball rolled forward. Our adder's snout appeared, and her head. Her tongue flickered. We stared silently. Unknotting herself in one long fluid movement, she veered to the left, turned back to the right and continued on into the heather.

Phil caught our Adder; freckly Phil, with steady eyes, bushy brown hair and large hands. I can see his face. Girls were always asking about him. But he stayed shy. In his eighteenth year he died, his motorbike ploughing into a pile of roadworks left unlit one night. A sports-car driver, the only witness, denied that he and Phil had been racing. 'Of course they were,' said Adrian.

Wild creatures are out in the open. Few live more than a few seasons. Predators swoop, ending life in an instant. An accidental wound can be a death sentence. Will we see that one again? Safe at home, we think of the animal, still out there. A snake is gripped by a hawk, or gets away. An animal goes under the wheels or escapes them. Moments ago a consciousness was there. It isn't now. The fish in another fish's mouth looks out as if from a hiding-place, the fins on the cheeks gently beating. Then the big fish gulps. The toad half-swallowed in the snake's jaws has a calmly attentive expression. Its golden eye shuts and opens, re-engaging with the world.

As I think of this, another example of my hesitation comes to mind.

It was a July night. I was twenty-two, just back from university. I had been out drinking with friends, and was reading in bed after midnight when the doorbell rang.

Immediately I felt deep alarm. More than alarm – I knew something very bad had happened. It was as if, somehow, I had been waiting for the sound. The house was silent. Perhaps because I was slightly drunk, I looked all around me, noting the things in the room, with a feeling that I was gathering myself to face the worst: a slightly giddy sensation.

No one else was up. I went downstairs, with a feeling of holding myself tall.

Two policemen were on the doorstep. They asked my name. Was my mother in? Could they step inside? I led them into the sitting-room. They told me they had bad news. What was it? My father had been involved in an accident earlier that evening, and it was fatal. I knew what the word meant, but it seemed an incidental word, as if they had said 'unusual'.

'Oh', I said, like a short sigh.

There was some sort of pause.

They were very sorry.

Did I need to sit down? No.

Would I go up and wake my mother? It would be better if I were the one to tell her. One of us would have to go with them to the hospital, to identify the body.

I went upstairs, holding myself straight, putting myself into the role. My life had become a novel or play. I was a young man about to tell his mother his father has died. But, as she struggled to wake up, blinking at me, I blurted out the news like a little boy. 'Oh, Mum,' I said. 'Dad's had an accident.' And, quickly, before she could get the wrong idea: 'He's been killed.'

I had said it in that simple sentence.

My mother, like me, seemed to understand at once and then backtrack into not grasping it.

Up in the attic rooms, a door opened and slammed; then another. My sisters appeared, open-mouthed.

The police drove us to the hospital, and the nurse said one of us should go in to the body. Perhaps I should go in, as it might be too distressing for my mother. I said I would go in.

My mother said, 'I want to go in.'

'In that case,' the nurse said to me, 'you wait out here. There's no need for both of you to go through it. Your mother will need you to support her afterwards.'

I was passive, grateful to be told what to do. Only on the way home, in the police car, did I think how I could have seen him, said goodbye, said sorry, said I loved him, reassured him, asked forgiveness, given him mine, stroked his face and loved him. I had let the moment pass. My failure to be there seemed like an abandonment – as if I had turned away and let him slip into darkness. But it was just that I didn't realise; didn't recognise the chance for what it was.

'It was him,' said my mother. 'But it didn't look like him.'

We went home to cry for weeks.

Later that night, I went up with my mother to their bedroom, clasping her hand. We had all taken something to help us sleep. In those days, most men slept in pyjamas, often striped pyjamas with a cord. My father didn't. He wore a red pair of nylon shorts; trunks, he called them. They were there, folded for him, on his side of the bed. 'Oh,' said my mother. 'Look, there are his trunks,' and she picked them up and kissed them. 'What shall I do with them?'

'Shall I take them away?'

'No,' she said. 'No. Leave them here.' She dropped them on the bed. 'Tomorrow I'll put them away.'

Much later that night I woke and wondered whether I could get up and go back to the hospital. He was still there, in that room. It might not be too late. He wasn't far away.

Chapter Six

Natterjack Toad, Aesculapian Snake

Sandscale Haws nature reserve is an area of sand dunes and beach in Cumbria, in the estuary of the River Duddon, near Barrow-in-Furness, where nuclear submarines are put together. Across the estuary, the mountains of the Lake District recede into hazy sky, one sharp crag behind another, fainter and fainter. In the evening sun, the near slopes are golden brown. They seem so close that I imagine I could touch one, reaching over the water to put my hand on the warm grass or feel the looseness of the scree. Beneath the mountains is a line of sand, glinting water, and more sand, the Sandscale beach. Turn away from them, and see towards Barrow an expanse of tussocky grass, hawthorn scrub, red and black cattle and, in the distance, chimneys with white plumes near pylons and long concrete buildings. The red cattle catch the sun too.

I have come here to see Natterjack Toads, an amphibian that, like the Great Crested Newt, we read about during our reptile craze, but never saw. It is not so surprising that we never saw this one. The toad specialises in sandy landscapes, and by then was scarce and highly localised in Britain. A hundred years ago, however, Natterjacks could still be found in London, and not long before that, they were widespread in the city. Thomas Bell, in 1839, observed the Natterjack to be common on Blackheath and numerous in ponds and ditches near Deptford. Malcolm Smith, writing in 1951, remembered finding Natterjacks as a boy at Coombe Hill near Putney and at Norbiton nearby. These sites were lost to building, and, by the time of Smith's book, the nearest colony to London was at Woking in Surrey. In the 1890s, Natterjacks were said to be common on the Surrey heaths. Small boys collected them to sell, calling them 'Jar-Bobs', according to the novelist Sabine Baring-Gould. At that time these little toads were still to be found in sandy areas with ponds and ditches in many parts of the country, especially south-eastern England, East Anglia and Lincolnshire, North Wales, the Lancashire and Cumbria coasts, and the saltmarshes of Solway Firth in Scotland. They also live in the bogs of County Kerry in south-western Ireland – the country's only toad.

Now, says Trevor Beebee, these toads have gone from about 75 per cent of the sites they had in the 1890s – for the usual reason: habitat destruction, due mostly to building and quarrying. Several sites in southern England lost their Natterjacks in the 1980s and 1990s because the ponds became too acid. The toad is now confined to a small number of duneland sites, such as Sandscale Haws, and one or two places on inland heath or moor. In Cumbria, there is a large population in a disused ironworks, living in the slag heaps. There is also a small mountain colony, quite isolated. Natterjacks breed mainly in shallow ponds where Common Toads and Frogs are not present, since the tadpoles of the common species hatch earlier and are already large and voracious when the Natterjacks are born. Outcompeted for food, the Natterjack toads do not grow.

The restriction to sandy soils is due to the burrowing habits of Natterjacks, and the difficulty this small running toad has in getting through dense vegetation. They hide or escape sun and frost by digging themselves into the sand with their back feet, and outside the breeding season they are often seen caked with sand. To hibernate, they dig deeper, sometimes scores of them together. As with the Sand Lizard, there is an ambitious programme of reintro- duction on sites where the habitat is safe. One of the most successful reintroduced colonies is at Hengistbury Head, a promontory east of Bournemouth, entirely cut off by the town.

Ten years ago, I looked for them there with my mother. After my father's death, she worked on in her London school for six years, before retiring to Swanage. We bought a house there with the compensation money, reduced because he had been over the alcohol limit. I was visiting.

'It's twenty-six years now since Dad died,' she said. 'That's longer than we were married. I was a different person then.'

Her voice was puzzled.

'It was a different life.'

The Bournemouth traffic was slow, and by the time we reached the reserve, there was only an hour of light left. We pottered around on the clifftop heath, finding nothing. The ponds gave no evidence of toads. A blustery wind came up, and it started to rain.

'Let's go back,' I said. 'You're getting wet.'

'What's that?'

'I said, let's go back.'

'No, listen. What's that sound?'

I heard it, through the wind, a faint high-pitched rasping. It came in surges, dying and starting again.

'It's them,' I said. 'It's the Natterjack call. Over there.'

'Do you want to go and find them?' she asked. The evening was very murky now.

'Just twenty minutes. Do you mind? I'll take you to the car first.'

'I want to come,' she said.

We set off in the direction of the sound. It came to us like a weak radio signal, lost in the wind and then suddenly strong.

'We'll have to leave the path,' I said. 'It's dark and muddy. Let me take you back.'

'Stop fussing. We're going.'

We stumbled over the lumpy heath. The call got louder. Now it was filling the air, a long grating trill, like an old-fashioned football fan's rattle. Several toads were calling, from different directions. My mother was somewhere behind me.

'Are you all right?'

'I'm fine.'

Ahead, there seemed to be a ditch. I shone my torch into it, to see a pair of eyes glinting up at me, a small toad in shallow water. My mother came up and put her hand on my shoulder. It wasn't a ditch, but a rectangular pond, newly made of black plastic, and covered with wire netting. There was sand on the bottom. Lines of spawn floated on the water, in curling patterns. A line of Natterjack spawn contains a single thread of black *vitelli*, not two or three threads side by side, as in Common Toad spawn. The *vitelli* are held in a line of slightly denser jelly, visible inside the larger string. Clusters of bubbles floated with the spawn.

At the touch of my torchlight, the male had stopped calling. I moved the beam away, hoping he would restart. Flashing it across the dark fields beyond, where other toads were still calling, I glimpsed tussocks, then similar ponds, about ten feet apart, all with wire netting covers. The other toads stopped. I switched off the torch. Mum and I stood there, perhaps for five minutes. One or two distant toads started again, but not ours. I switched on the light. Where the skin of his throat had ballooned out, it was shrinking and collapsing, like bubble gum, stretched thin to white transparency.

'Dad would have liked that,' said my mother, back in the car. 'He would have loved doing it with you.'

'Oh, he wasn't really interested, was he?'

'Well, it wasn't his thing. He was more into history and politics. But he wanted to do things with you. I wish you'd done more with him.'

'He got so angry with me. He was hitting me all the time.'

'I know. He was very hard on you. We all thought that. It was partly his nerves, you know. He was always very sorry.'

I started the car.

'Do you remember that time he took you out to catch lizards? You and those friends of yours, Adrian and Philip. And did Micro go? No, you had split up with Micro.'

'We went to Hayes Common, yes.'

'Dad told me he enjoyed that day so much. You really wanted a lizard, and hadn't caught any when it was time to go, and he found you one, and he waited for you, keeping an eye on it, and you watched it together. He knew how much you wanted it. But you said you wouldn't try to catch it, because it was pregnant.'

'Yes,' I said. 'I don't know what stopped me that time. Plenty of other times I caught pregnant females.'

'Not that time. You told Dad you wouldn't. Perhaps you decided not to catch it because he was there.'

'He told you all this? I thought he barely noticed what I was doing.'

'He told me. He told his friends. He told people at work. He was proud of you, Richard.'

'I knew that,' I said.

It was true. In the fiercest of our battles I don't think I ever felt other than confident that he loved me. I could hurt him because I knew that – I *wanted* to hurt him because I knew that. It was a way of trying to break through.

There was one great mercy. At about the age of eighteen, I stopped fighting with my father – certainly stopped fighting in the old way. A large part of it was that I was now doing the things he wanted. I was going to go to university; I was fascinated by literature, history, politics; I was a precocious sixth-former, full of dreamy ideas. He liked that, and he and I began to feel the strange sensation of enjoying each other's company.

Of course, he could still be infuriating. There was one party I was at with a girlfriend, and I had told him we would come back to sleep at our house, where she always slept in the guest room. The time came to leave, if we were to catch the last train. But we didn't want to tear ourselves away. It was embarrassing to say we had to leave; embarrassing to me, anyway. I'm not sure that Alison minded. I was heady with the experience of being at a party, with girls, music, cannabis, intellectual talk. It was still a kind of miracle to me that I had arrived at all that.

So I phoned him, on my friend's parents' phone. All around me, people were talking, drinking, dancing. Down the corridor, I could see they were kissing.

'Is that you, Rich?'

'Yes – look, we're still at the party. We've missed the last train. Don't worry; Simon says we can stay overnight. Lots of people are doing it. We'll kip on the sofa or something. It'll be rooms full of people.'

'Is Alison still there?'

'Yes. She's fine with it.'

'Well, you can't do it, Rich. It's not right.'

'What?'

'It's not right.'

'Oh. Look, Dad, really.'

'What would Alison's parents say? You told them she was staying the night *here*.'

'They wouldn't mind. It's only you that reacts like this.'

I tried to say this in a relaxed conversational tone, as if discussing some piece of gossip. It came out half-stifled.

'I'll come and get you in the car.'

Because of the people around me, I was dumb. Half of me was trying to think of arguments to put to him, half full of rage at him for embarrassing me, and half already resigned to leaving the party: three halves.

I made a shamefaced explanation to Simon.

But this was a throwback. In those last three years or so, I felt a strange flooding fondness for him. We argued politics. He told me how impressed he was with the left-wing young teachers at his school, how they made him feel he had always been so timid. We teased each other. It was extraordinary for me. We went to the pub, something we had never done together. I found myself using certain tones of speech he had, tones I thought of as derived from his Suffolk childhood. Those tones gave me a sense of safety, of family laughter and home.

The Natterjack Toad is a small short-limbed toad, algae-green or sandy brown. Their moist pimply skin in the water has the colour and look of floating algae with bubbles trapped beneath it – green but slightly rusty. Darker patches of red and green cover the surface. The toad's belly is creamy white, with some brown or grey marbling, the surface composed of tiny bumps or bubbles. Down the middle of the back runs a pale yellow line, vivid and straight, except for the odd little kink, as if a painter's hand had trembled. I did not understand how such a conspicuous mark could have evolved on such a perfectly camouflaged animal, until, at Sandscale Haws, I saw what the line mimics – floating stalks of straw, from reed or marram. Now I notice such stalks floating on every pond I visit. They are always there. Pool Frogs and Edible Frogs have this line also.

Camouflage often works not only by mimicking the colours and markings of plant, stone or tree trunk, so that the animal melts into the background. It also works positively, by offering a different outline from that of the animal: tricking the eye into seeing a different shape. If you see the line and it registers as something, a straw, then your eye thinks its work is done. You don't look for the toad. Perhaps the Grass Snake's yellow collar works in the same way, making the glancing eye think 'buttercup'.

No one seems to know the origin or meaning of the name 'Natterjack'. 'Of obscure formation' says the *Oxford English Dictionary*, suggesting a possible combination of 'attor', from the

Old Norse 'Eitr', a poisonous liquid that was also the origin of life, and the common name Jack, meaning a fellow or upstart. Thus, 'Natterjack' could mean 'poisonous fellow'. This is a very tentative theory. Alternatively 'Natterjack' could have something to do with the German word for snake, 'Natter'. Our 'adder' comes from the older word 'nadder' (apparently 'a nadder' was frequently misheard and written 'an adder'), deriving from the Old English *naedre*, meaning 'nether' or 'lower', which had the same Saxon and High German origins as 'Natter'. It is hard to see why a snake should be invoked in the naming of this toad, though Grass Snakes certainly eat them, but the theory that both 'Natter' and 'Natterjack' come from *naedre* seems plausible. Reptiles and amphibians were traditionally seen as 'lowly' creatures in the moral sense, as well as creatures living low down on the earth.

The Victorian naturalist Mordecai Cubitt Cooke, writing in 1865, suggested a combination of *naedre* and the German word *jäger*, meaning hunter or runner. According to this theory, 'Natterjack' means 'low runner' or 'crouching runner', and the word has sometimes been used as a verb, naming an awkward style of movement, half-running, half-hopping. This usage does not explain the origin, however. More likely, it derives from the name already used for the toad.

It is true that these animals have a distinctive style of movement. They run, rather than hop like a Common Toad or a frog, and they are good at it; surprisingly fast. J. E. Taylor, another Victorian observer, says that they run 'almost like a mouse'. A local name for Natterjacks, when they were common, was 'Running Toad'; another, recorded by Taylor in 1884, was 'Walking Toad'. After all this, I am glad to say that the scientific name has long been uncontested. The Natterjack Toad is *Bufo calamita*, the name assigned by Josephus Nicolaus Laurenti in 1768. To me, as a child, this name always suggested 'calamity', and it was strange to think of this endearing and eccentric little creature as some sort of bringer of biblical disaster, or sufferer from it. In fact, the name comes from *calamus*, a reed. The Natterjack is the toad of the reeds.

People know that they live near these toads because of the call, the repetitive undulating rasp, like a fingernail on a comb. In the long breeding season, which can begin in March and continue to June or July, living near Natterjacks is like living in cicada country. On a warm evening, as scores of males join in, trying to be loudest, the sound can be deafening. In Surrey, these toads were called Thursley Thrushes. On Merseyside, they were Birkdale Nightingales. The male makes the call by forcing air rapidly from his lungs into the expanding vocal sac that balloons out from his throat, and then sucking it back. At full stretch, the sac becomes almost the size of his body. A group of us gathered at Sandscale Haws, to hear this performance.

The year had started badly for Natterjacks. Drought, in January, February and March 2012, had dried up the shallow slacks and scrapes at many sites. At Ainsdale Dunes, near Southport on the Lancashire coast, one of the most important Natterjack sites in the country, there had been no breeding for three years running. The Sandscale warden explained this to us as we waited. In these circumstances, he said, Natterjacks do not breed at all. If the pools have dried up before the toads come out of hibernation, the scent and moisture triggers are not there to bring the toads into breeding condition. This is natural resilience. The survival of the colony is not endangered by a year or two like this. Four or five years is another matter. 'This year makes it three, at Ainsdale and Formby,' the warden said. The threat is serious because Natterjacks in Britain are in a predicament shared by all but the commonest British reptiles and amphibians. Their colonies are isolated. Sand Lizards are in this position, except perhaps in Purbeck, and even Adders, still widespread across the country, are disappearing locally for this reason. Without habitat corridors between different sites, colonies cannot replenish naturally after disasters, and gene pools cannot be refreshed.

Sandscale is one of these isolated sites. So is Hengistbury. These places are surrounded by urban development and water. But Sandscale is managed primarily for Natterjacks. There are few competing priorities. Ponds have been designed and constructed for the toads, and kept topped up. Another factor is that, at Sandscale, some of the spawning takes place in mud shallows at the estuary edge, where the water is still mainly fresh. This is where fresh water from the Duddon joins the sea. If the seawater content is no more than 15 per cent, Natterjack spawn and tadpoles will survive. These shallows do not dry out in the absence of rain, though they are vulnerable to storms and tidal changes. In places, the tadpole shoals blacken the shallows.

It was a dull evening now. Wind was blowing the marram and throwing up sand. Rain was spitting. We were sitting beside the largest pool, but had heard no calls. Blankets were laid out; folding chairs opened. Tupperware boxes came out. A couple poured me a coffee from their thermos. 'Last week, we counted thirty males calling here,' said the warden. 'But it was warmer then. Don't worry, though,' he said. 'We have our secret weapon.'

He produced a portable CD player.

'On here,' he said. 'I have a recording. We made it a couple of years back, on one of the loudest nights we've ever known. There are males in this pond, I'm sure of that, and females are still arriving. Tonight's a bit cold and windy, that's all. But once they get going, it'll be fine. We need something to get them interested.'

He pressed a switch. After a bit of static, the call rasped out. He turned up the volume, let us listen for a moment and switched it off.

Meanwhile, another warden was wading in the pond. To handle a Natterjack toad, or disturb it or its habitat, one needs a license from Natural England. Only the wardens could wade. The rest of us sat on our blankets. Everything I would have done as a child was forbidden.

'There's one here,' she called.

'A male?'

'A nice one.'

'I'll bring the player.'

He switched it on, and the call trilled out over the shadowy dunes, making us look up. An aircraft flashed its landing lights, moving towards Barrow. On the main road, headlights appeared as a car turned a corner. The call got louder, coming in pulses.

'I think he's starting,' called the warden in the pond. 'His throat's puffing out.'

'Just a few minutes more.'

The quiet was a shock, but a thin rapid sound was still there, a reedy piping. It carried on for a minute, and stopped. A gust of wind troubled the water. The sound started again, for a pulse or two, stopped. We waited, and looked at each other. Abruptly, a much louder trill began, soaring up to a drawn-out crescendo, with a slowing fall of single notes, dry and throaty. Silence followed. There was a tentative clap from someone. We all joined in, laughing.

'I want it for my ringtone,' said a young woman.

'Thank you,' said the warden. 'Thank you, ladies and gentlemen. I'm not sure we'll get an encore, but I can invite you to meet the soloist. Can you catch him, Kathy?'

'I think so.'

She waded out of the water, and presented him to us, a tiny, moist greeny-red toad, shrinking down on her palm in the torchlight. He gathered himself, swallowed hard and looked up. The patterns on his back were like green lichens on a rock. Reddish bubbles dotted his surface. His yellow line was bright. Cameras were flashing.

Why do these animals still excite me, and comfort me? Every year I go looking for them, in my old haunts at Studland, and in new places – Adders in Kielder Forest among the first primroses, Natterjacks at Sandscale and lime-green Sand Lizards at Birkdale. On the islands of Lundy and Flat Holm in the Bristol Channel, I look for Slow Worms, finding gnarled old specimens larger than any I have seen elsewhere, and males so densely marked with flecks of blue that they appear to be made of blue jasper, and surely come from some Mediterranean temple, not windy England. I am determined to see those native Pool Frogs in Norfolk.

This summer I went with my friend, the artist Hugh Nicholson, to the banks of Regent's Canal, near Camden Lock, opposite London Zoo. We were looking for the Camden Creature. Hugh had seen it the week before, quite by chance, and had agreed to show me the place. Eating his lunch on a bench by the canal, looking over the water to the zoo, he had heard a rustle, and turned to see something like an electricity cable lying between two clumps of brambles. It moved, and he saw that the cable was a snake, a long one, unlike any of the British snakes. For a few minutes he watched the creature thread its way through the stems and the trash of takeaway cartons, beer bottles and cans, tongue flickering. Unhurriedly, it moved into the brambles.

The Aesculapian Snake, *Elaphe longissima*, is a lithe, smooth, shiny brown snake that can grow to more than six feet. Its head is slim, long and flat on top, its nose abruptly blunt, its eye round and pronounced. The brown gives way to greenish yellow on the cheeks and belly; these are the only two colours. This is a tree-climbing snake, which you may see in the branches above you. Aesculapian Snakes are common across southern Europe, from most of France and the very north of Spain, through Italy and south-eastern Europe to south west Ukraine, with isolated populations in southern Germany and southern Poland. They feed on small mammals, especially rodents, and are entirely harmless to people.

Persistent reports began in the 1990s of strange snakes seen on the overgrown canal embankment, on both sides, near the zoo. Photographs identified the species. Estimates put the population at between twenty and thirty-six. Between 2007 and 2010, one observer, watching casually but frequently, saw seventeen individuals. They cross in and out of the zoo. Rats and mice on the canal banks are a plentiful food supply. In many of the photographs – which can be seen on the Reptiles and Amphibians of the UK Live Forum (search for 'The Camden Creature') – the snakes have large bulges. Part of the embankment is south-facing, and on the bank there is a mulch of vegetation, composting down and generating warmth. It is likely that this is where the snakes lay their eggs, the same size as a hen's.

How the colony got there is a mystery. Three possibilities have been suggested – that they escaped or were released from the zoo, that they somehow came from the famous Palmer's Pet Store a few streets away – where my friends and I used to gaze longingly at chameleons, iguanas, and baby alligators – or that they either escaped from a study centre for schools run in the 1980s by the Inner London Education Authority or were released when the centre closed. Aesculapian Snakes were frequently sold in pet shops until the late 1980s. A population like this could derive from just one or two individuals. There is another feral colony of this species in Britain, which seems to have similar origins – they live inside and just outside Colwyn Bay Zoo on the North Wales coast, and the twenty or so snakes are apparently descended from one pregnant female that escaped in the 1960s.

Since news of their presence got around, the Camden Aesculapians have become an attraction for British reptile-lovers, but they are listed by the Department for Environment, Food and Rural Affairs as an invasive species that could cause harm to native wildlife. It is hard to see how, since the Camden population is entirely cut off from any native reptiles, and not in an area where any rarities need protection, while the Welsh population has not expanded beyond one very small locality in nearly fifty years. Nevertheless, it is possible that an eradication programme will be put into effect. In 2013, the Wales Biodiversity Partnership, a body that reports to DEFRA, recommended that the Colwyn Bay snakes be removed from the wild, arguing that the zoo should be responsible for carrying out the plan. Reptile enthusiasts have suggested angrily that this is an example of a government agency 'ticking boxes' by selecting easy targets. They point out that no action is proposed to eradicate feral pheasants, refugees from the shooting estates, which seem likely to be killing a lot of reptiles. Evidence of this is at present anecdotal, but plentiful. Wyre Forest in Worcestershire, for example, seems to be losing the population of Adders that the famous 'citizen scientist' Sylvia Sheldon has so lovingly studied and monitored. Some observers blame the marked increase in pheasant numbers. The shooting industry, however, is a formidable force to confront. See the Camden Creature while you can. It may not be there much longer.

Hugh and I did not see it, though the August day was sultry. We sidled quietly up to clearings where a snake might bask and prodded the thick vegetation. A man in a blanket, with a bottle of wine, eyed us curiously. There was a wasp nest in the mossy bank, with insects flicking in and out. Across the water, a pair of warthogs rooted in their dusty enclosure. Up the bank, out of sight of the path, we could see syringes with needles and the ashes of campfires.

The question of alien species is difficult. They can do great damage. Across the world so many species are now listed as critically endangered. The International Union for Conservation of Nature – the organisation, supported by more than two hundred governments, that monitors the extinction crisis, compiling the Red Lists of critically endangered, endangered and vulnerable species – identifies Invasive Alien Species as a global cause of extinction and endangerment second only to Habitat Loss and Degradation. New arrivals can have quite unexpected consequences. The emergency is a real one. It frightens conservationists and zoologists everywhere. It frightens me.

Amphibians, says the IUCN, are the most threatened group of species, their permeable skins and their movement between land and water making them exceptionally vulnerable to skin infections and ecological change. Nearly a third of the world's amphibian species are in danger of extinction. The latest particular threat comes from a fungal infection called *Batrachochytrium dendrobatidis*, or 'chytrid', which has recently spread rapidly across the world. Chytrid thickens the animal's skin. A severe attack prevents the absorption of water on which amphibians depend. Mass deaths have occurred in many places, though they do not always occur when the fungus is present, and scientists are not yet sure why. A dangerous moment in the amphibian life-cycle is the moment of transition, when the tadpole becomes a frog, toad or newt and is ready to leave the water. In Spain, herpetologists have peered over the edges of ponds and found thousands of corpses beneath overhanging banks. Chytrid is certainly spreading in Britain. It was first detected here in 2004, and though few animals have died here so far, there is great concern, and amphibians should not be moved from one pond to another. How the disease got into Britain is a mystery, but imported species that have escaped or been released, such as European Alpine Newts and African Clawed Toads, are prime suspects.

269

Talk of 'aliens' bringing contamination can sound like an echo of sinister political forces in the human world, but such comparisons do not go very far. There is no necessary connection between wanting to preserve biodiversity from threats brought by non-native species (however shifting the definitions of 'native' may be) and racist attitudes to other human beings. Conservationists have to write rules, and write them in the context of a global crisis that could take away much of our animal life for ever.

Ecological consequences often confound predictions. Nevertheless, it is truly hard to imagine any harm that these isolated London snakes could do, and I like to think of them twined in the trees here, and in the moment of their appearance turning the quiet canal bank, with its familiar urban wildness, to a jungle that can make your neck prickle.

Something attracts us when we are young, and enters our imagination, becoming the doorway to adventure. Reptiles and amphibians gave me sensations of discovery, sensations of hunting, sensations of the world as infinite, leading us on. We develop our personal symbolism, our poetry. What attracts a lot of boys is the fantasy land of war. Aggression and defiance find a symbolism there. One can feel exultant. I experienced this when I read those comics and daydreamed about myself as a war hero. Wild nature offers symbolism too. The attraction of bird-watching must at least in part be the apparent freedom of birds – the open spaces they inhabit and their ability to take off and go. The contrast is with everything that holds us down.

With reptiles, especially snakes, part of the attraction for me was the drama of seeing the animal suddenly – the abrupt grip of excitement, driving out every other thought. They were creatures of dramatic entrances. Important, therefore, was my perception of snakes as an outcast species, the enemy of everything approved of by my father, my teachers and the popular children at school. I didn't want to *be* the snake, exactly. Two ideas – more than ideas, visions and fantasies – were huge in my mind. There was forgiveness, the idea of the second chance that would not be messed-up – the idea that everything I had done wrong would be understood, and I would be welcomed back and loved. And there was the idea of anger, of unassuaged feeling that had to be recognised, had to be expressed and given space, before forgiveness and homecoming could be accepted. I identified the snake with this anger, this bitter vengeful enmity, that really wants to make friends, but doesn't know how.

What I needed to do, with the powerful masculine animals that excited me, was come to see their weakness, as well as their power. The bull is led into the slaughterhouse, trembling. I saw a bull once in a pen in a busy livestock market. His panting had whisked his dribble into a ball of froth that covered his whole muzzle. The snake loops helplessly in the hawk's claws or my hands, or lies mashed on

271

the road. My father trembles inside the tank, inside that fetid metal box, expecting the blast.

If something was your passion in childhood, if it became your symbolic language and the ground of your battles, then that is a reason to return to it throughout your life. It brings contact with your past – a reassurance your life is still there. I have been out to look for reptiles in many states of emotion. Once it was the after-shock of bereavement – not in the first few weeks after Dad's death, which were a blur of tears, all of us crying together, but about six weeks later. A friend invited me for a day's walk, on the North Downs. I was speechless for much of the walk, striding on in a daze. But we came to a place that seemed familiar. Had I been there before? I am not sure. But I recognised that it was prime lizard country, a hollow with bracken and gorse, and a large rotten log. 'There might be lizards here,' I said, and immediately saw one, twenty yards off, running along the tree trunk. I walked over. It was a female Common Lizard. And there was a baby, on a yellow tussock, just at my feet. My tears came again.

I have gone out to watch them during the feverish beginnings of falling in love, when I wanted to shout with excitement, but was afraid, too, and wanted something to calm me. I remember a time when I did not hesitate; not eventually, anyway. It was about a year after my father's death. A woman and I were walking in the fens near Cambridge. I had been falling for her for several months, but did not know whether she felt similarly towards me. It was an April day, changing between warm sun and windy showers. I had invited her for a walk in the fenland landscape, hoping to move her by showing her some of the wild nature I cared about, hoping also for the release of my own emotions, enough to bring me to say something or make a move.

But by mid-afternoon, the tension had not yet broken. We had seen lapwings tumbling in the air and giving abrupt shrieking cries, and kingfishers, and a greyish bird in the distance that might have been a Marsh Harrier. In the black water of the drainage ditches, we had seen several pike, hanging in the water, their fins paddling slowly. When the sun came out, shafts of light in the water revealed soft floating particles. One of the pike shot forward and turned so sharply that its body was forced onto its side; we did not see what it was chasing. I hoped to see Great Crested Newts, but these drainage ditches were deep and dark, and it was difficult to get close. The soil here is black, and against the mud on the ditch bottom, any newts would have been almost invisible. I did point out a couple of toads on the surface of the water, and began to tell some of the story of my interest in them. We were talking pleasantly, confidingly – but I didn't know how to ask the question, make the overture. I was afraid of the embarrassment if my approach was unwelcome. We walked side by side along the ditch in silence now, looking straight ahead. I felt oppressed. Something had to happen. What was she feeling? Was it a boring walk for her?

On a tussock ahead of us, I saw a Grass Snake coiled tight – a large one, shiny from the spring's shedding. Its face, with the tiger stripes descending from the eye, pointed up at us. The tongue appeared, flickered, went in.

'Look,' I said, touching her arm.

She swung towards me, responding to my touch more than I expected.

'There – in the grass.'

'Oh yes.'

Her hand slipped around my arm.

We crouched to look at the snake, which slowly uncoiled and let itself gently down to the water's edge, not really alarmed, tongue exploring the air. The snake slid into the water, and set off into the middle.

Our two bodies, shoulders touching, were leaning out over the edge. With her arm still on mine, I let myself sit back, on the top of the bank, and then slid my legs down, so that my feet rested near the water's edge. I lay back against the bank. She did the same.

I leaned over, and my mouth touched her cheek.

She turned her face to meet mine, her mouth open and ready, and we let ourselves slide down the bank, without breaking the kiss, until we almost went in, and had to stop, and laugh, and brush ourselves, and scramble up to where we knew we would start again.

Reptiles and amphibians in Britain are in many ways in trouble. Populations of the common species are declining. Rare species are only preserved by intervention. Habitat loss is the greatest disaster, and reptile enthusiasts often feel furious that even the nature-loving organisations, such as Natural England and the RSPB, forget reptiles when it comes to land-management. Sometimes, heavy equipment is used to clear ground vegetation and reptiles are wiped out in the process. Individuals are killed. Cover is removed. Hibernacula are destroyed. Immediately after such a clearance, reptiles are sometimes seen more easily, but this is because of the loss of cover, not because they have not suffered, and in the open these creatures are vulnerable. Vegetation returns, but the reptiles may not. Small isolated colonies may not recover.

There is reason to worry, then, and complicated work to be done, if these curious creatures, disturbing, endearing and challenging to our imagination, are to remain a common presence in wild Britain, and therefore in our culture as it responds to natural wildness. I hope they will. In this book I have tried to show why. In my life, these animals have played an important part – a strange one, perhaps, and an accidental one, but one that has continued. When something has happened to me – something to shock me, grieve me or delight me – I have always, after a while, gone walking on the heath to look into its hiding places, or stopped the car by a pond and walked over to gaze in, or spotted a piece of flat metal on the grass to lift up. My eye looks around until it comes to some sort of wild margin. I always find one. That's what I did on my first day at that new school. It is what I did preparing for university exams. I have done it when falling in love, and when desperate after a love has been lost.

When I came back to London as a teacher, I climbed over the fence into a park one night, in New Cross, to shine my torch into a pond, and see what was there: Smooth and Palmate Newts. 'What you doin', bro?' said a voice. I looked up, and saw a ring of teenage boys around me. 'Looking for newts,' I said. 'See that one there?' In my torch beam a leopard-skinned male cruised through the water, shimmying gently. 'Wick-id,' said one of the boys. 'You a naturalist or something?' And they all gathered round.

Arriving in a place, or taking leave of it, I look around for those margins: the overgrown empty housing lots, the railway banks, the inaccessible riverbanks, the roadside banks, the edges of the runway. From train windows, I see woods, lakes, mountain slopes, deserts, woodland clearings that dance with butterflies. 'What might be there?' I wonder, every time.

Further reading about
British reptiles and amphibians

If you would like to encounter these animals and find out more about them, here are some good sources of information.

WEBSITES

Reptiles and Amphibians of the UK (RAUK)
(www.herpetofauna.co.uk)

The site has a live forum with a section on each of the British species and on topical questions about conservation. Specialist scientists, professional conservation workers and ordinary enthusiasts come together in lively debates, and people upload pictures of the reptiles they have seen. The first animals to emerge each year are always greeted with enthusiasm. A lot of anger is expressed about loss of habitat, and policies on nature reserves that are damaging to reptiles and amphibians. This site is friendly, funny and informative: ideal for finding out where and when to see reptiles and amphibians, and how to understand what one sees (though because of the fear of collectors, people are often wary of giving precise locations).

Amphibian and Reptile Conservation (ARC)
(www.arc-trust.org/)
This is another highly informative site – a noticeboard for professionals and volunteers working in conservation. ARC is a national wildlife charity devoted to the conservation of herpetofauna. It organises scientific and conservation projects, and public awareness and rescue campaigns, and owns more than eighty small nature reserves with important reptile and amphibian populations. Details of current projects can be found here, including calls for volunteers. The site also provides summaries of relevant legislation, details of conservation methods and equipment, and lots of legal and scientific links. There is an online shop selling reptile-related merchandise.

The British Herpetological Society (BHS)
(http://thebhs.org/)
The BHS is the country's main scientific association for the study of reptiles and amphibians, founded in 1947 and currently possessing a membership of more than 600. Membership brings access to three journals – *The Herpetological Journal*, which carries academic research papers, *The Herpetological Bulletin*, which publishes scientific articles more suitable for a general readership as well as letters, book reviews, answers to queries, and summaries of herpetological news,

and a more informal newsletter, *The Natterjack*. Members also have access to a specialist library, and to meetings including the December scientific meeting now run jointly by BHS and ARC – a meeting whose purpose is to communicate current research. BHS also funds the purchase of land for nature reserves in consortium with ARC and local Wildlife Trusts, and sponsors research projects in Britain and overseas.

The Surrey Amphibian and Reptile Group (SARG)
(www.surrey-arg.org.uk)
SARG is a prominent county group for enthusiasts, with one of the best reptile and amphibian websites. Activities include training events for conservation professionals and volunteers, and the co-ordination of teams of volunteers rescuing toads from traffic at migration times. Two excellent and distinctive resources on this website are a bibliography of British historical literature about reptiles and amphibians (www.surrey-arg.org.uk/SARG/15000-Books/SARGhistoricBooks. asp), and a link to the pages of the Wall Lizard Project, a survey run jointly with RAUK of all the identified sites where Wall Lizards can be found or have recently been recorded in Britain. An entry for each site provides, with a map, such details as are known of the colony's history and a recent estimate of numbers (www.surrey-arg.org.uk / SARG/02000-Activities/SurveyAndMonitoring/WallLizard/ PMSitePicker.asp).

European Field Herping Community
(www.euroherp.com/)

This is a website devoted to 'field herping' – the finding and photographing of reptiles and amphibians in the wild. Field herping characteristically involves some disturbance of the wild animals. Logs, rocks and other forms of cover will be quickly turned over – this is called 'flipping' – and the animals revealed will be captured by hand and posed for photographs before they are released. Field herping of this kind cannot be practised on the heavily protected species in Britain without a license. Contributors regularly upload detailed narratives of field herping expeditions, mostly in continental Europe. Many of these reports are fascinatingly eventful and richly illustrated with photographs. The site sometimes goes inactive for lengthy periods and has several times been flooded with spam, but always available, come what may, is the wonderful archive of photographs of all the European species. Similar sites, with a mainly continental focus, are www.lacerta.de and http://fieldherping.eu/

Books

In *Cold Blood* I have quoted frequently from two historical publications, Edward Topsell's *Historie of Serpentes*, published in 1608, and Thomas Bell's *A History of British Reptiles*, published in 1839. Topsell's book is a compendium of descriptions, beliefs, rumours and anecdotes from a wide variety of ancient sources. Bell's is the first scientific natural history book on the British species, revealing a great deal about how widespread they were and how they were perceived. Some of the woodcuts that illustrate Bell's *History* have been reproduced in *Cold Blood*.

A delightful book that is now rare is *The Lives of British Lizards* by the poet and herpetological scientist Colin Simms, published in 1970. It has a tone unlike that of any other scientific natural history book I know. Mainly, Simms restricts himself to scientific information, but there is a subdued lyricism all the way through the book that very occasionally surfaces to become more explicit. Of his personal story he provides very little, but his love of the animals is clear.

There is only one contender for recognition as the most substantial and influential twentieth century book for the general reader. Malcolm Smith's *The British Reptiles and Amphibians*, a work in the famous Collins New Naturalist series, was published in 1951. Written with beautiful clarity, and far more detailed than any previous popular work, it remains a good source of information and is still fairly easily obtainable from second-hand bookshops or from the

Advanced Book Exchange (ABE) online. New Naturalist volumes are prized by collectors, and some are now rare and hideously expensive, but this one is quite common still. The jacket designs by Clifford and Rosemary Ellis are famous in their own right, and on this book the chalky semi-stylised picture of the back of an Adder moving through long grass has become very familiar to me. It makes the spine stand out on a shelf of second-hand books. Spotting it is a bit like spotting a real Adder.

In *The New Naturalists*, his history of the series, Peter Marren explains that in the 1940s new cameras and the availability of Kodachrome film made high-quality photographs of living small animals a possibility for the first time. Corpses or stuffed animals had previously been used for the photographic illustrations of natural history books. I did not know this when, as described in *Cold Blood*, I first opened the pages of Edward Step's *Animal Life of the British Isles*. A few of the pictured animals were obviously dead. The others looked alive but were stiff in posture and staring wildly. I found these pictures disturbing and powerful.

The Kodachrome colour plates in Smith's book were powerful in a quite different way. I saw them for the first time in the library of my new school, and the impressiveness of the book became associated in my mind with the intimidating atmosphere of the old-fashioned high-ceilinged reading room. Only the quietest whispers were permitted and, often, the prefect on library duty would cough fiercely to hush even those. Dark wooden shelves went up far above my height. A staircase on wheels stood in waiting. Smith's book, with the section of Adder on the spine, happened to be at eye-level.

On the first plate I saw was a female Great Crested Newt, poised vertically in the water, arms outstretched, with her belly towards the camera. She looked as if she was letting herself sink slowly. The yellow was intense, and the white of the sand on the bottom. Individual grains of sand were sharply clear, and so was the warty texture of her body. All the colour plates had this intensity and stillness – this combination of bright but natural looking colour and unusual clarity. In this first plate the power was increased by the fact that I had never seen a Great Crested Newt in the flesh, but in every plate the effect made the animal appear to be an exceptionally fine specimen: one that if you saw it in the wild would stop your breath and make you think, 'I've got to catch it.' The Common Frog was a lovely red, the Common Toad a complex map of greens and browns with each wart clearly defined. A scatter of baby Slow Worms on a rock were polished gold. Only the Natterjack Toad looked wrong – it was too sugary a green.

I also associated this book with the images I had then of science, especially biology – images that came from my glimpses of labs at the new school. Science was something that happened on dark wooden benches, not the white plastic surfaces I think of now. The biology lab had cork-lined display cabinets and cupboards of animals preserved in bottles, squeezed in and contorted, and losing their colour. It was rumoured that we might be given live frogs to dissect. This never happened at my school, but dissecting frogs in biology lessons was still a common practice, and a friend of mine tells a story of how his frog suddenly started to move. They were imported from India by the thousand. Smith's book had secondary illustrations also, black and white pictures of animals with their eyes closed that looked as if they had been taken out of laboratory bottles for the purpose. You would not see such pictures in a work of popular natural history now.

Comparing Smith's book with the equivalent works today, one sees how the emphasis in science has changed. An up to date New Naturalist volume on herpetofauna by Deryk Frazer was published in 1983, and a successor by Trevor Beebee and Richard Griffiths in 2000. This third book, *Amphibians and Reptiles*, is now the best substantial work on these animals for anyone interested in finding out more. This readable, well-illustrated book will provide a thorough grounding quickly. As Beebee and Griffiths point out in their introduction, they place less emphasis than Smith on anatomy and much more on ecology and conservation methods, partly because of the immense development of ecological science that has occurred since Smith's time, but also because of the severity and urgency of our new environmental problems. It is poignant to compare Beebee and Griffiths with Smith, or even with Simms, because the comparison in each case shows us how greatly the range of these animals in Britain has diminished – and how they could once be taken for granted as familiar wild creatures, part of a shared public experience. In the 1940s, when the New Naturalist series started, the spirit that impelled it was a widespread feeling that Britain's wildlife was an important part of the nation's popular heritage – the home to which the troops would be returning, and the reward for the sacrifices of the war. This spirit is very strong in the early books in the series, and in other natural history books of the time. It is still present in the more recent books, but we could do with a rediscovery of that idea, reshaped for our time.

An excellent pocket guide to the British species, which is also a good introduction to their ecology and conservation, is *Britain's Reptiles and Amphibians* by Howard Inns, published in 2009 (it must be harder and harder to find a new slight variation on the only basic title these books can have). The colour photographs in this book are the best I have seen in a recent volume. In a light, slim book (with a waterproof cover), Inns manages to include pictures that show a lot of the colour variations. A handbook with more detail of current ecological questions and projects, and methods of survey and monitoring, is Trevor Beebee's *Amphibians and Reptiles*, which has a very useful chapter on 'How Schools can Help'.

For the continental European species as well as the British, there are two good field guides. The one most easily found in bookshops is *Reptiles and Amphibians of Britain and Europe* in the Collins Field Guide series, written by Nicholas Arnold and Denys Ovenden. First published in 1978, and updated in 2002, this guide has distribution maps and good coloured drawings of numerous species and sub-species, helpfully laid out. A similar, slightly smaller guide with good photographic illustrations is Axel Kwet's *European Reptile and Amphibian Guide*, published in 2009.

BOOKS MENTIONED

Edward Topsell, *The historie of serpents. Or, the second booke of liuing creatures wherein is contained their diuine, naturall, and morall descriptions, with their lively figures. Names, conditions, kindes and natures of all venemous beasts.* First published 1608. A paperback reproduction was published by EEBO Editions, ProQuest, in 2010.

Thomas Bell, *A History of British Reptiles* (John Van Voorst, 1839).

Colin Simms, *The Lives of British Lizards* (Goose & Son, 1970).

Malcolm Smith, *The British Reptiles and Amphibians* (Collins New Naturalist, 1951).

Deryk Frazer, *Reptiles and Amphibians in Britain* (Collins New Naturalist, 1983).

Trevor Beebee and Richard Griffiths, *Amphibians and Reptiles* (HarperCollins, 2000).

Howard Inns, *Britain's Reptiles and Amphibians* (Wild Guides, 2009).

Trevor Beebee, *Amphibians and Reptiles* (Pelagic Publishing, 2013).

Nicholas Arnold and Denys Ovenden, *Collins Field Guide to the Reptiles and Amphibians of Britain and Europe* (Collins, 2004).

Axel Kwet, *European Reptile and Amphibian Guide* (New Holland, 2009).

Acknowledgements

Will Francis, my literary agent, saw the potential of this book early on, and encouraged and helped me as the idea took shape. When I began to put short pieces of writing together, he was an exacting reader, with an unerring eye for phrases that were not clear. For that and for his skill at marketing the proposal and negotiating on my behalf, I give him deeply-felt thanks. Clara Farmer has been a superb editor, very precise, full of understanding of what I was doing and expert at helping me get to the finish. I owe her a very great deal. Susannah Otter steered me expertly through my nervous last edits, understanding exactly what was needed. James Jones designed a cover that caught the spirit of the book more exactly than I would have thought possible. My thanks go to all of the team at Chatto and Windus.

Chris Nicholson, Tessa Hadley, Greg Garrard, Paul Evans and Tim Liardet have read sections of the book and given expert and encouraging advice, and, more importantly, have supported me with humour and honest, steadfast friendship. Chris, especially, found many literary leads for me and gave me the frankest of criticism. Kitty (Catharine) Nicholson's astonishing drawings of woodland and heathland micro-landscapes have been an inspiration to my attempts to write about such places. SueEllen Campbell taught me a great deal about how to integrate scientific and personal writing. For twenty years I have benefited enormously from being able to work with students and teachers on the MA in Creative Writing at Bath Spa University. In so many ways that experience has helped me break through into the writing I always longed for. Students there, and tutors such as Richard Francis, Steve May, Colin Edwards, Gerard Woodward and Jeremy Hooker, have shown me so much of what writing involves, and some of it sank in.

Fiona Sampson responded with sharp and sympathetic under-standing to a section of the work that she heard me read aloud, and subsequently published part of it in *Poetry Review* – support that was crucial (volume 101:4, winter 2011). An early version of another part appeared in *Granta Online* (2 December 2011: http://www.granta.com/New-Writing/Our-Adder). Chris Davis, a leader of the Sand Lizard Recovery Programme, gave me an interview that was warm, funny and generous with information. Cate Sandilands gave permission for the quotation in Chapter Two from her memoir/essay 'Landscape, Memory, and Forgetting' (published in *Material Feminisms*, edited by Stacy Alaimo and Susan Hekman, Indiana

University Press, 2008). She also commented usefully on *Cold Blood* in response to a reading I gave at the conference of the UK and Ireland branch of the Association for the Study of Literature and Environment at the University of Surrey in 2013. ASLE, in both the UK and the USA, has been a steady source of relevant discussion. I have benefited also from the annual scientific meetings organised by the British Herpetological Society, and from passionate debates on *Reptiles and Amphibians of the UK*, the website set up by Chris Davis (http://www.herpetofauna.co.uk): especially the contributions of Gemma Fairchild and Wolfgang Wüster.

Trips to hot reptile and amphibian spots across the UK were supported by Bath Spa University and by the Roger Deakin Award for 2012, given by the Society of Authors. Rees Cox and Helen Fearnley gave me stories about Smooth Snakes and Sand Lizards. Hugh Nicholson took me looking for snakes along Regent's Canal. Kate and Robert Rigby were with me on two occasions when I seemed mysteriously to have conjured snakes, one of which I have described in the book. Paul and Shelly allowed me to sit at their dinner table writing frantically for a week, and did their best to help me look for rattlesnakes on the hillside at the bottom of their garden, though I am not sure they hoped as much as I did for success.

Thanks to my parents Paul and Dorothy and my sisters Cathy and Ann. They did not ask to be characters in this book, and my mother and sisters have responded with generous feeling. I hope they see *Cold Blood* as the loving book I believe it to be. Thanks also to all the friends who had reptile and amphibian adventures with me.

Most of all – in a home more dominated by matters reptilian than anyone would have chosen but me – thanks to my daughters Lily and Violet for their love and for so much happiness, to Imogen who once came on a reptile expedition at the age of six and screamed when an Adder crossed the path inches from her feet, and to Tracy, for reading chapters in successive versions and giving detailed advice every time. And for her love, patience and belief.

www.vintage-books.co.uk